A Microscopic Analysis of Use-Wear and Polish Formation on Experimental Quartz Tools

Carole Sussman

BAR International Series 395
1988

B.A.R.

5, Centremead, Osney Mead, Oxford OX2 0DQ, England.

BAR -S395, 1988: 'A Microscopic Analysis of Use-Wear and Polish on Experimental Quartz Tools'

ISBN 9780860545101 paperback

ISBN 9781407346731 e-book

DOI https://doi.org/10.30861/9780860545101

A catalogue record for this book is available from the British Library

This book is available at www.barpublishing.com

For Glynn Llywelyn Isaac

Table of Contents

List of Figures

Figure

List of Tables

Table

List of Plates

Plate

Plate

Preface

This work, as a Ph.D. dissertation, was aided by a multitude of people both at the University of California, Berkeley and Harvard University. Changes to the content original dissertation are minimal with the exception of minor editing and the addition of new information on microwear studies gained at the Use-Wear Conference held in Valbonne, France in 1986.

The purpose of this monograph is to provide an easy reference guide for anyone attempting microwear analysis on stone tools made of quartz. It should be emphasized that this is the result of preliminary research and is, by no means, a definitive work on the subject. Ongoing research by myself and others, will undoubtedly shed new insights on quartz microwear and the field of use wear analysis as a whole.

When this work began in 1983, little attention was paid to the effects of discard and burial on microwear polish. Recent work has emphasized the need to consider the effects of soil movement and soil acids, as they can alter and even obliterate use wear polish. It is my belief that microwear analysis is a valuable research tool in discerning past behavior through stone tool use but, it is possible that after we tease out the effects of thousands of years of burial, the remaining information may be less precise than originally hoped. Still, great advances are often made up of many small steps and I do not feel, as some, that microwear should be discarded as a useless endeavor.

Many people gave freely of their time and ideas to this work, and I am grateful to all of them for their help. Any misinterpretations or errors within this monograph are solely my own.

J. Desmond Clark provided training and support both in the classroom and in the field. His ongoing interest in new methods and ideas always offered fresh prospectives for my research. This monograph is, in no small way, due to his encouragement.

F. Clark Howell and Garniss Curtis read various drafts of the thesis and made constructive suggestions and comments. Andrea Lain worked tirelessly on numerous experiments. David Pilbeam and Jack Harris are also to be thanked for their time, help and ongoing encouragement.

Sincere thanks goes to the following people for their technical advice and assistance: Jochim Hampel of the Earth Sciences Department at U.C. Berkeley; Madeline Anderson, Allen Sonafrank and Gene Hammel of the Quantitative Anthropology Lab at U.C. Berkeley and Maurice Dumais, John Merkel and Dave Lange of Harvard University.

The staff of the Anthropology Department at U.C. Berkeley, was particularly helpful in various aspects of my research and writing, especially M.J. Tyler, Lillian Tamblyn and Charlene Hunter. The Anthropology Department at Harvard University and the staff of the Peabody Museum, also generously facilitated aspects of this research.

My colleagues at the Stone Age Lab at Harvard University influenced the content of this thesis, and I am grateful to Jeanne Sept, Nikki Stern, Annie Vincent, John Shea, and Greg Laden for constantly sharing new ideas and friendship. Peter Jones made some of the tools I used in my experiments, and was most generous with his time, help and advice. Jim Moore convinced me that statistics were the 'way to go' and provided unlimited access to a computer. Thank you all.

The late Glynn Isaac first proposed this topic for my dissertation research; he also gave me the opportunity to go to Harvard and set up a microwear lab where I could carry out this research. The many kindnesses and generosity expressed by him and Barbara Isaac will not be forgotten.

Glynn's stimulating ideas and enthusiasm for this project sometimes surpassed my own. He always found the time to discuss and guide all aspects of my research. His positive outlook, boundless energy, and creative thinking are sorely missed.

Financial assistance for my research was generously granted by the L.S.B. Leakey Foundation, the Foundation for Research into the Origins of Man, a University of California Humanities Research Grant, Robert H. Lowie scholarships, and the Boise Fund. I would like to express my gratitude to all of them.

Chapter 1

Introduction

Through the 1960's, man was considered as separate from the Great Apes due to his ability to make and use tools (Bordes 1968; Clark, G. 1969, 1970; Clark, W.E.L. 1962; Oakley 1956). Studies of free-ranging chimpanzees (Pan troglodytes) by Jane Goodall and others (Goodall 1964, 1968; Boesch and Boesch 1981, 1983; McGrew et al 1979; Nishida and Vehara 1980; Sugiyama and Koman 1979) showed that apes regularly adapt and modify objects in their environment for use as tools. Chimpanzees strip twigs and stems of their leaves as for use as termiting sticks, crushed green leaves are sponges to sop up water for drinking, and rocks are use to crack open nuts. While other animals such as baboons, birds and sea otters are also known to use tools (Beck 1980; Guilmet 1977), as far as we know, hominids became the only apes who habitually made and used durable tools - those made of stone.

It is likely that early hominids also employed branches, bark and leaves as tools, but, being organic, such tools are generally not preserved in the archaeological record. This forces present-day anthropologists to rely on analogies with our closest living relatives, the chimps, for comparison (Zihlman 1979; Zihlman et al 1978; Susman 1980). Inferential evidence from the early site of Koobi Fora in Kenya dating to 1.5 million years before the present indicates that some of these early stone tools were being used to work wood, perhaps for making digging sticks, to cut soft plant material, and to cut meat as well (Keeley and Toth 1981). It was this ability to make and use stone tools that gave the comparatively slow hominid (Lovejoy 1981; Susman et al 1985:190) an edge over the other animals competing for the same resources.

The first use of stone tools signified an adaptive shift from bipedal hominids who probably modified branches and leaves, to utilizing sharp-edged flakes that could be made by striking, with preconceived intent, one stone against another. These could be re-used, re-sharpened, and were more efficient in performing such tasks as processing an animal carcass and rendering it into manageable packages of protein (Jones 1980; Leakey, L.S.B. 1950; Sussman, 1985) or digging in the earth to expose roots and tubers and, perhaps even more importantly, using flakes as tools to make other tools - digging sticks, for example (Vincent 1985).

Historically, the first documented recognition and study of stone tools in the Sixteenth and Seventeenth centuries raised questions concerning how these were made and where they came from. They were described as weapons of a primitive people ignorant of metals; their true antiquity only hinted at, and were also claimed to have been generated in the sky, being a mixture of thunder and lightening mixed with "metallic matter..." (Daniel 1975:25-26). The discovery of handaxes in association with the bones of extinct animals raised a strong argument for accepting these stone tools as the handiwork of a prehistoric people (Ibid:26). This strengthened the proof for man's antiquity, placing it earlier than 4004 B.C, the date which Bishop Ussher derived from the Bible for the creation of the world (Daniel

1975; Wood 1976). The actual functions of these stone tools were not questioned or well examined until later in the Nineteenth century by curious, inquiring people such as Spurrell (1892).

Conventionally, stone tools were classified into typological categories based on their overall morphology and inferred function. During the last part of the Nineteenth century and the first half of this century, archaeologists directed their attention and efforts towards more systematic excavation of sites and amassing collections of shaped stone tools while discarding, for the most part, unretouched flakes and other waste (Daniel 1967). These collections were used as "markers" to date and construct hierarchical sequences based upon their comparative morphology, which were then used to explain population movements and/or cultural change (Braidwood 1946; Childe 1925; Laville et al 1980; Sackett 1981; Trustees of the British Museum 1968). The development of absolute dating techniques by Libby et. al. (1949), freed archaeologists from the necessity of using stone tools for dating and allowed attention to focus on other relevant issues such as tool function and assemblage variability. Function was originally deduced from the shape of the stone tools, the angles of the edges, and comparisons with ethnographic data (See for example, Binford, S. 1968 ; Burkitt 1963; Clark, J.D. 1958; Gould et al 1971; Kamminga 1977; Mason 1967:744; Nissen and Dittemore 1974; Odell 1981; Rigaud 1978; Summers 1950; Tindale 1965).

The variation found in lithic assemblages was attributed to natural diversity, activity variants, stylistic components, "culture" or a function of the raw material (Binford and Binford 1966; Binford 1973; Bordes and de Sonneville-Bordes 1970; Clark and Kleindienst 1974; Clark, J.D. 1980; Close 1978; Dunnell 1978; Jones 1979; Kleindienst 1967; Laville et al 1980; Leakey 1950a; Mellars 1970; Rigaud 1978; Sackett 1973, 1982; Tucker 1980). Consequently, up through the early 1960's, numerous typologies and terminologies arose to classify and define the different assemblages, relying heavily on artefact morphology, composition and some quantitative analysis (Bourlon and Bouyssonie 1912; Balout 1955; Bordes 1950, 1961; Goodwin 1958; Isaac 1967; Kleindienst 1961; Leakey 1970; Mason 1962; Movius 1948; Sackett 1966; Tixier 1963).

At the Burg Wartenstein Symposium in 1965, a concerted effort was made to standardize lithic terminology and analyses of artefact assemblages by adopting comprehensive terms to eradicate some of the confusion, vagueries and controversy arising from the use of numerous terms and typologies (Clark, J.D. 1967; Daniels 1967; Mason 1967; Kleindienst 1967; Summers 1967; Tixier 1967). While these controversies have not been completely eliminated, it is clear that strong headway has been made.

An experimental approach towards the study of stone tools began to take hold in the early part of this century. Important contributions have been made by many people, among them Bordes (1947; Bordes and Crabtree 1969) and Crabtree (1972), two accomplished flintknappers and pioneers in the field, who studied and experimented with the mechanics of stone fracture as well as recognizing the need to examine flintknapping debitage in order to gain a comprehensive understanding of the reduction sequences (see also Tixier 1972; Tixier et al 1980).

Another interesting approach has been taken by Jones (1979) for assessing the practical effects of stone tool 'blanks' and edge sharpness by using different raw materials. Jones replicated bifaces from Olduvai Gorge in Tanzania and demonstrated the differences in edge sharpness before and after resharpening, as well as the limits placed on tool morphology due to choice of raw material by early hominids (see also Toth 1982).

Early hominid behavior has been a tantilizing unknown that archaeologists have sought to understand in a variety of ways: through studying the location and character of sites (see Isaac 1968, 1971, 1978 for examples), by exploring spatial relationships amongst artefacts on the sites (Cahen et al 1979), examining the remains of bones and other features within sites (Binford 1981; Cahen and Keeley 1980; Clark and Haynes 1970; Shipman et al 1983), conjoining studies of bones and stones (Bunn et al 1980; Cahen and Keeley 1980), and microscopic examination of stone tools and associated bones for traces of modification and utilization (Anderson-Gerfaud 1983; Bunn 1981; Keeley 1977; Moss 1983a,b; Potts and Shipman 1981; Shipman and Rose 1983; Sussman 1985b).

One valuable approach that has received much attention is the technique of "use wear" analysis using microscopy to examine the edges of stone tools for traces of use (for an in depth review refer to Cook and Dumont 1983; and to Odell 1983 for pros and cons of both high and low power microscopy). A piece of stone which has been used as a tool may have one, or a combination of the following features along the edge which was utilized: edge damage in the form of small scars of various types; striations from use; abrasion or polish. These attributes have been documented and described, using a multitude of techniques, to determine exactly how tools were used and the material(s) they were used on. A brief summary of the various types of use wear analyses is given below.

The "Father of Use-Wear" is a title often used to describe Sergei Semenov (1974). While not the first person to use microscopy for analysis of stone tools (Curwin 1930), Semenov was the first to systematically research and document use wear traces on the edges of stone tools. This work has been continued and expanded upon by others (Clark and Prince 1978; Kamminga 1978; Kantman 1970; Phillipson and Phillipson 1970; Tringham et al 1974; inter alia). The thrust of their work has been to quantify and describe the scars left along the working edges of stone tools using magnifications generally less than 100X (also referred to as "low power" use wear studies). Depending upon the types of scars found, their size, patterning and distribution, statements can be made regarding the type of material the tool was used on, e.g. "hard" versus "soft", as well as the action employed. Critics of this approach (see Tomenchuk 1983:73; also Odell 1983:17) argue that it fails to provide specific information about the type of material a tool was used on, the terms "hard" and "soft" being seen as inclusive and rather vague. It has also been established that edge damage caused by other factors such as burial or trampling can closely mimic retouch or utilization and create "artificial" tools (Flenniken and Haggarty 1979; Gero 1978; Moss

1983c).

Further work in the 1970's using high power (>100X) incident light microscopy identified distinct polishes which were left on the edges of flint tools after use (Keeley 1974; Keeley and Newcomer 1977). While polish such as "sickle gloss" was already known and noted by Spurrell as early as 1892, Keeley (1980) found that other polishes, distinctive to the material worked, also exist; so a tool used to scrape dry hide, for example, sustains a very different polish from a tool used on bone. This technique, known as microwear analysis, also works well on other fine-grained materials such as chalcedony (Beyries 1982; J. Guysels pers.comm.; author pers. obs.) and ignimbrite (Keeley and Toth 1981). For the first time, scientists are provided with the opportunity to gather more specific information about the behaviour surrounding these tools - what functions they had and the activities performed by the hominids who used them.

Microwear studies thus aid morphological analyses and may even shed a new light on the continuing 'style versus function' debates by determining upper boundaries of stylistic variability before the functional purpose of a tool must change. Like low power use wear studies, however, this technique is not without its critics who contend the subjective nature of the observations and rightly argue that the results should be more objective and standardized (Odell 1975). Competency of microwear analysts should also be tested by administering rigorous "blind" tests (refer to Cook and Dumont 1983; Newcomer et. al 1986).

Microwear analysis is not appropriate for use on all raw materials, such as coarse-grained rocks like basalt (Sussman pers. obs. but see Plisson 1982 for opposing view), or obsidian (Hay 1977; Vaughan 1981). Microwear polish is affected, and can be obliterated by natural processes such as abrasion by wind, water and soil movement, (Shackley 1974; Yamada and Shimura 1984), chemical alteration of the stone by weathering (Fojud and Kobusiewicz 1982), as well as "bag damage" from collective storage (Keeley 1980). With all its shortcomings, microwear analysis has provided archaeologists with the most specific functional information on stone tools to date.

Recent work by Levi-Sala (1986) has produced preliminary findings which stir the winds of caution. She has altered experimental use wear polishes artificially in the lab by tumbling both used and unused flint tools in sand to mimic post-depositional soil movement. It is not suggested that artificial tumbling mimics the effects of tens or hundreds of thousands of years of post depositional action and soil acids but, her work serves as a reminder that we have, as yet, merely begun to scratch the surface in functional use wear studies and that the effects of burial on microwear polish are more pronounced than originally anticipated. It is possible that microwear specialists may find themselves relying on the mechanical features produced by wear to identify use, such as striations and grooves, with polish attributes playing a more supportive role. Important research such as this is an example of the new direction use wear studies are taking, where the emphasis is now on unravelling and identifying the mechanical and chemical processes affecting polish after discard (see also Plisson and Mauger 1987).

This monograph is the result of in depth experimental research attempting to expand the applicability of microwear analysis to include experimental tools made of quartz, a stone that was widely used in Africa and Asia during the Lower Palaeolithic. Kjiel Knutsson, of the University of Uppsala in Sweden, is the other researcher who has recently concentrated his efforts on microwear analysis of quartz tools. The research presented here was conducted independently of Mr. Knutsson's work but, recent discussion with him has shown that our results corroborate each other's work.

Quartz is abundant in the archaeological record, spanning from the earliest known stone tools from the Omo in Ethiopia (2.3 m.y.a., Merrick et al 1973; Merrick 1976; Merrick and Merrick 1976), from Zhoukoudian in China (500,000 B.P., see Lanpo 1980) through Paleo-Indian times in the New World. Expanding microwear analysis to include quartz will allow scientists to study artefacts from important sites such as the Omo, Olduvai Gorge (Leakey 1971; Sussman 1985b), Border Cave (Beaumont 1973; Beaumont et al 1978), inter alia.

In general, the research design consisted of striking flakes from large quartz crystals, and then using these flakes on various materials for specific lengths of time to document the presence and development of both edge damage and use wear polish. The materials worked were those known, or thought to have been used by early hominids for particular activities and included: bone, wood, hide, meat, shell, and antler. The activities included butchering dead animals for meat and perhaps skin and sinew, cutting and whittling branches to make pointed sticks for digging, sawing and graving bone and antler, and cutting soft plant material, perhaps to use as bedding or ground covering. Activities are discussed in their appropriae sections. After use, all the tools were cleaned and examined microscopically. Flakes of Brandon flint, a homogeneous dark grey flint from England, were also incorporated into the experiments. Polish attributes first described by Keeley (1974) were on stone tools made from Brandon flint, so using tools made from it alongside the experimental quartz tools provided an excellent means of comparing the "classic" microwear polishes with the wear patterns and polish found on quartz.

I want to emphasize that the aim of this research project was twofold: 1) to establish if use wear polish is a visible and distinctive entity on quartz tools and 2) if the polishes left after use are identifiable to a specific material (such as hide versus bone) on experimental quartz tools. Future research will include testing these results on selected quartz tools from appropriate archaeological assemblages. Another aspect of future research includes expanding the technique to encompass a larger sample of vein quartz. Vein quartz, the milky-white crystalline variety, was used more often in the making of prehistoric stone tools than crystal quartz. This is presumably because vein quartz is relatively abundant the world over, whereas large quantities of crystal quartz are relatively rare. Vein quartz is more difficult to examine microscopically due to its highly reflective and uneven surface structure. The attributes of use-wear polishes are, however, the same for vein quartz but, are much more difficult to recognize, unless well developed, for the following reasons:

1) the small crystals of quartz do not "wear" at the same rate along the working edge of the tool due, most probably, to the uneveness of the microtopography; confident identification of polish, therefore, can be difficult, especially if it is not well developed.

2) the unretouched edge of a flake of vein quartz tends to be more friable than that of an unretouched flake of crystal quartz, and breaks more readily during use.

3) the uneveness of the microtopography on a tool made from vein quartz makes focusing at high magnification extremely difficult.

Chapter 2

Methods and Terminology

Raw Material

Quartz is a brilliant, hexagonal crystalline mineral made of silicon dioxide, which occurs abundantly in colorless, transparent form although, it may also be colored by impurities as it is forming. Crystals of quartz may vary as to their symmetry (Figure 1) and flaws or the beginning growths of other, smaller crystals are commonly visible inside the main crystal. It has a hardness of 7 on the MOH scale, a specific gravity of 2.65 and lacks cleavage, often fracturing conchoidally. When a substance fractures conchoidallly, a hertzian cone of percussion forms upon impact, as the force of the blow spreads and dissipates through the mass. This principle is well illustrated when a beebee pellet hits a pane of glass and the force created forms rings of percussion which subsequently punch out a tiny cone. Chemically quartz is extremely stable and endures the ravages of time with little alteration.

To determine if differences in molecular packing on the various axes of a quartz crystal would affect polish attributes, I took the advice of one of my advisors (G.I.) and included flakes obtained from different axes of the same crystal in some of the experiments . When used on identical materials for the same amount of time, polish attributes were the same for all the flakes, regardless of their crystal orientation. This finding was verified using a scanning electron microscope in association with the incident light microscope.

Archaeologically, crystal quartz is an excellent raw material for manufacturing stone tools; it fractures easily into sharp-edged flakes which can be used "as is" or further retouched into a variety of forms. Many stone industries in Africa (Clark 1950; Jeffreys 1951; Leakey 1971; Merrick 1976; Miller 1969), Asia (Lanpo 1980), as well as in Australia (Flood 1983; McCarthy 1976) and in the New World have a percentage of their tools made from either crystal and and/or vein quartz.

More abundant and, therefore, used more often for stone tools is vein quartz; a variety of quartz found as nodules and cobbles. It differs from crystal quartz being micro-crystalline in structure and usually milky white. Vein quartz does not fracture conchoidally, but, more often than not, breaks into irregular chunks and formless flakes along faults and crystals, making it extremely difficult to shape into "formal" tools. There is, however, vein quartz found in tabular blocks and large boulders at places like Naibor Soit, (the quartz outcrop near Olduvai Gorge in Tanzania where early hominids obtained some of their raw material for tools) which is more easily fashioned into large, well-shaped tools such as bifaces (P.Jones pers.comm.). Very fine lanceolate points were also made from vein quartz during the Middle Stone Age (J.D. Clark, pers. comm.).

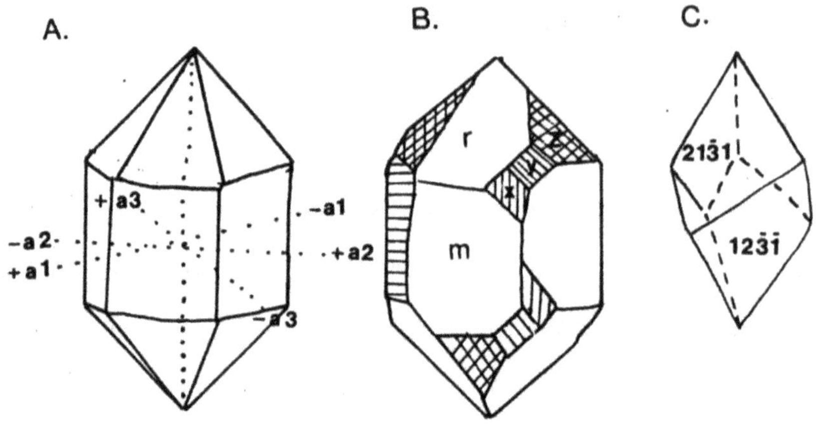

QUARTZ CRYSTALS

A. Crystal of common type showing highest hexagonal symmetry

B. More complex revealing true symmetry. The forms represented are the prism m (1010); rhombohedron r (1011); rhombohderon z (0111); trigonal trapezohedron x (5161) and trigonal pyramid y (1121)

C. Trigonal trapezohedron (2131) of quartz showing true symmetry

(adapted from Read & Watson 1966:30)

Fig. 1

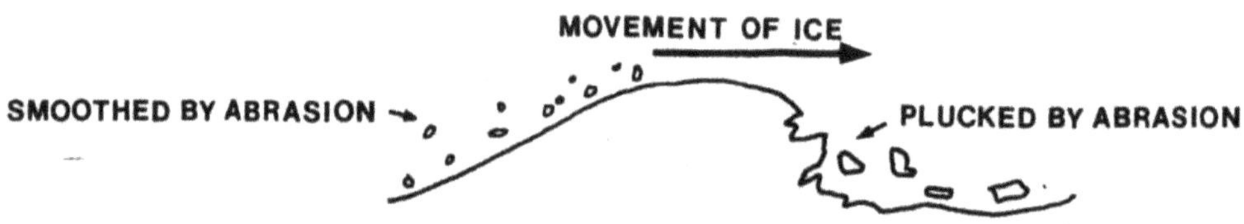

GLACIAL MOVEMENT

(adapted from Mears 1978:300)

Fig. 2

Microscopically, both vein quartz and crystal quartz are extremely reflective and variable in surface configuration as well as surface texture. Since vein quartz is made up of many small crystals, when viewed microscopically, reflectivity is increased tremendously, making identification of features, such as wear patterns, extremely difficult. The microtopography of a flake of vein quartz is generally much more irregular than a flake of crystal quartz, making focusing an arduous, and sometimes impossible, task. Plates 1 - 4 are representative of some of the variability found on unused quartz surfaces.

For these reasons, I used crystal quartz for my experimental flakes. I felt that once I was able to recognize and define the attributes of use wear on quartz, it would be easier and more accurate to examine more difficult material such as vein quartz. I did use a few pieces of vein quartz early on in the experiments and these will be discussed in their appropriate sections.

Procedure

Flakes of both quartz and flint were made by direct percussion using a hammerstone and were either caught in the hand or allowed to drop to either a tarpaulin or the grass below. Quartz flakes (generally those to be use for the longest amount of time) were examined microscopically and photographed prior to use to document: a) any irregularities on the surface of the quartz which could be misinterpreted as "wear" and b) the presence of micro-scarring and edge damage sustained during manufacture (refer, for example, to Plate 4).

The edges of the flakes were left unretouched whenever possible, allowing use damage to determine the morphology of the edge. Occasionally, for certain tasks such as scraping, a steeper edge angle was required for efficiency, so a hammerstone or an antler baton was used to retouch and strengthen the edge.

All the flakes were assigned a number and both the dorsal and ventral sides were drawn to scale on a sheet of metric graph paper. This allowed accurate drawing of changes in the profile of the edge from use, location of polish on the tool and any pertinent remarks.

Each flake was used for its assigned length of time to perform a specific activity such as: hide-scraping, cutting soft plant material, wood working, etc. Both flint and quartz flakes, when used together, worked the same materials for the same amount of time. After I had a representative collection of experimental flint tools to compare with the used quartz tools, I discontinued using them and concentrated instead, on building up the sample size of used quartz tools. All these tasks were carried out, whenever feasible, in natural settings to replicate as closely as possible aboriginal conditions. No additional grit or dirt was added (except in one unusual hide scraping experiment described later), nor was any attempt made to remove grit from the experimental environment. Cutting meat for butchery experiments, for example, was done by placing the meat on top of the hide itself or on top of a wooden table. It was never placed directly upon the ground.

Time was chosen as a variable rather than the number of strokes although both methods have been used (see Keeley 1980; and also Tringham et.al 1974). Ethnographically, tasks that are time consuming such as hide scraping or bow and arrow making, are generally carried out in a relaxed fashion, often as part of a social group, whereas short tasks such as making digging sticks, for example, are done quite swiftly - often in a matter of minutes (Silberbauer 1981; Yellen 1977; Vincent pers. comm). Therefore, the number of strokes required to perform a particular task to completion will vary undoubtedly as will the amount of pressure exerted on the tool as it is used. The angle at which the tool is held will also vary a few degrees in either direction during the course of the activity. Using time as the independent variable was an attempt to circumvent the "artificiality" of the experiments and approach more "normal" or realistic conditions. Because this research dealt with human, hence variable, activity (as opposed to using a use wear machine), the regression and discriminant analysis I used for a portion of the analysis are not statistically valid due to the unavoidable (and unmeasurable) variation in work force and effort used by the experimentors (G. Hammel pers.comm.). Nonetheless, the results are quite interesting and useful and should not be discounted entirely. They serve to indicate trends in attribute developments and are valuable in that respect.

Various materials were used in the experiments: hide (fresh, sun-dried and tanned); fresh bone; antler (dry and soaked); meat (raw); vegetal matter (grasses); wood (fresh and aged of various species); feathers (fresh) and shell. The results of each of these "sets" of experiments will be discussed separately. After being used, all flakes were cleaned in the following manner:

1. a ten minute soak in ammoniated detergent to remove any fingerprints. The tool was rinsed under running water for one minute before being placed in,

2. a 10% solution of HCl for ten minutes to remove any mineral deposits which might be adhering. While this is not a problem with experimental tools, it is often necessary when viewing archaeological specimens. The HCl can also be heated if necessary when deposits are stubborn and the soaked period prolonged. Again, this is followed by a one minute rinse with clean running water, then,

3. a ten minute soak in a 10% solution of KOH to remove any organic deposits. This is followed by another one minute rinse with running water and,

4. a ten minute soak in clean ammoniated detergent again as a final "insurance" followed by another one minute rinse with clean water to remove traces of detergent.

5. Drying was done quickly and briskly wearing rubber gloves and using either lint free cotton cloths or a hair dryer. Gloves were worn while handling clean specimens because oil from the hands is readily deposited on the surface of the flakes and obscures wear attributes and polish.

I found no apparent alteration of the polish when the soak was prolonged, nor were any differences noted between those pieces cleaned using acetone compared with those undergoing chemical soaks. At present, however, I prefer to use acetone and water to remove dirt and finger grease when cleaning artefacts but, at the time these experiments were done, I adhered as closely as possible to Keeley's technique (Keeley 1980).

6. Each specimen was stored individually in a numbered paper envelope with a tracing of the tool on the outside. Individual storage of the tools prevents abrasion and scratching from contact with other pieces.

7. All specimens were examined a minimum of twice with the light microscope (Olypmpus BHM metallurgical microscope) to locate areas of damage and polish for description and photography and for location and measurement of striations. The most useful magnification was 200X both for locating areas of wear and description of wear attributes. Photographs were taken using an Olympus OM-2 camera with an automatic light meter and Agfapan 100 black and white film.

Selected pieces were studied with a scanning electron microscope (JEOL-35 and ISI DS-130) to detect surface features or textural changes which might otherwise be obscured by the light reflection under the metallurgical microscope. Using a SEM required that each tool be coated with approximately 300 angstroms of gold to make the surface conductive. Magnifications used with the SEM ranged from 10X for edge profiles to 2000X; with the most useful range being 200-1000X. Higher magnifications were used but not extensively due to the uneveness of the surface topography of the tools which made fine focusing quite difficult. Flakes photographed while inside the SEM were not tilted unless otherwise stated and were mounted on a stub so the edge of the flake was parallel to the stub. Duco cement and silver paint worked well for mounting the specimen to the stub securely and providing a good contact between the two. Care should be taken that the tool is absolutely free from finger grease and dust before placing it in the SEM so that extraneous material is not introduced onto the surface of the tool. The specimen was easily removed from the stub by a brief soak in pure acetone and then a thorough cleaning with ammoniated detergent and water.

The scanning electron microscope uses electrons to create an image in contrast to the reflected light used in the metallurgical microscope. Plates 6 and 7 illustrate some of the differences between the light microscope and the SEM. Plate 6 is a photo taken, using the light microscope, of pits and striae created by cutting grass with a quartz flake for 60 minutes. Using a magnification of 200X, the features are easily seen. Plate 7 is the same area taken at the same magnification with the SEM. Note that the arrow points to the same feature in both photos. It is more difficult to find these features in the SEM photo due to the lack of reflected light. Plate 8 is a photo of the same area under the SEM magnified 1000X. The features, which were readily visible using the light microscope (Plate 6), are more easily seen now. It is interesting that features, such as comet-tailed pits (Plate 15), which were readily visible under the light microscope at 200X, were detectable in the SEM only at extremely high magnifications and only when the specimens were tilted away from the electron gun. Unfortunately, tilting specimens in the SEM creates distortion of features as the electrons shooting from the gun strike the surface of the tool at oblique angles. The amount of distortion created increases proportionately with the degree of tilt, making it difficult to measure attributes precisely but, easier to detect extremely shallow surface features such as comet-tailed pits and striations with a tilt of approximately 30 degrees (Plate 9). These shallow features are easily detected using a metallurgical microscope because, the light reflection enhances them, making them more pronounced. Plates 10-13 illustrate the amount of distortion that occurs when a piece is tilted; the arrow in each photo points to the same feature. Notice how the feature becomes elongated as the angle of the tilt increases. The micron bar in each photograph has been corrected for abberation caused by the tilt.

While reflected light is problematic for viewing quartz microscopically, due to the irregularity of the surface and the quantity of light reflected back off the surface, the metallurgical microscope remains essential for recognizing the more subjective, optical qualities of polishes, such as the "satiny smoothness" of an antler polish or the "metallic brightness" of a silica polish (refer to Newcomer et al 1986 for problems associated with this approach). By eliminating light reflection and increasing the depth of field, the SEM is invaluable for isolating surface textures, in addition to studying edge damage and abrasive action which are often masked by surface irregularities (see also Anderson 1980,1981; Meeks et al 1982 for use of SEM on flint).

Much of what happens to the surface of a quartz tool is abrasive in nature similar, in many respects, to glacial movements (see also Kamminga 1979). Glaciers erode by abrasion caused when the debris-laden ice scrapes and rasps against the underlying rock. This smooths and polishes the surface of the rock, altering it into gently rounded forms. These mounds, ridden over by the ice, become streamlined and smoothed on the side facing the source of the ice while the downglacier or lee side of the mound becomes roughened and steep-faced (Figure 2). When a glacier is loaded with fine debris, the ice may polish the hard rock beneath it. Jagged-edged cobbles dragged over the underlying rocks make striations or grooves which parallel the direction of movement of the glacier (Andrew 1975; Brainerd 1978; Sanders et al 1976). These asymmetrical knobs of bedrock are called "roches moutonees" and are a common occurrence (on a much smaller scale) on quartz tools used to work moderately hard substances such as wood, bone and antler. Plate 14 illustrates the "roches moutonees" effect quite well on a quartz flake used to grave soaked antler. Another feature common to both glaciers and quartz is plucking, or the removal of fragments of rock at the base of the glacier or in low lying areas of quartz (Fig.2 and Plate 14).

As there are many terms which will be continually used throughout this monograph, it seems wise to define them now, thereby avoiding repetition in each section.

Definitions

Pits - Pitting, as defined by Diamond (1979), occurs when surface fatigue is created after repeated contact between two surfaces sliding over one another (see Plates 6 & 8). The term pit, refers to the depression left after microscopic fragments spall off the surface of the tool. There are also pits, according to Diamond, which are merely depressed areas on the surface of the tool that have become isolated from further abrasive action. While Diamond refers specifically to flint, I found the same features to be present on the surface of quartz tools and suggest that the phenomenon of spalling occurs on quartz as well. In this study, I did not note, however, pits corresponding to Diamond's depressed areas.

Keeley (1980:50) describes particular pits found specifically on hide-working tools as "pot-lid" fractures. They are circular in outline and are caused by heat generated from the friction between the hide and the flint tool during use. While pitting is an extrememly common feature on quartz hide-working tools, I did not observe "pot-lid" fractures of this type.

Comet-tailed Pits - These contrast with the pits described above due to the adition of shallow depressions or "tails" trailing after the pit. It is interesting that I found these comet-tailed pits **only** on those quartz tools used to work either fresh bone or soaked antler (Plate 15). They have been noted on flint tools used for cutting soft plant material (Keeley 1980; Witthoft 1967), however, I never found them on quartz tools used for cutting grass. On quartz tools, these pits are extremely shallow and easily visible using a light microscope at magnifications of 200X or greater. Finding them using the scanning electron microscope required magnifications of 500X or greater, and a tilt of approximately 30 degrees. Even then, they were difficult to see because they lacked depth (M. Dumais pers. comm; Kamminga 1979). As with other linear features such as striations and linear grooves, comet-tailed pits run parallel to the direction of use. The placement of their "heads", (e.g. the pit), is more variable, usually facing the working edge in the case of scraping, but sometimes away from it. This variability in direction seemed to be random in its occurrence although no statistical counts were made to test this assumption. It is possible that the "head" of the comet-tailed pit results from the initial contact between the quartz surface and the substrate, and the "tail" is formed as the velocity of the two materials to each other changed (Del Bene 1979; Keeley 1980; Witthoft 1967).

Linear Grooves - These are gouges or rough bottomed striations which also parallel the direction of use (Plate 16). They occur when a small fragment, such as a microchip of quartz, is pressed obliquely onto the surface of the tool during use, and maintains a force parallel to the quartz surface, overcoming contact friction. "At the trailing edge of the contact the specimen experiences an enhanced tension due to the drag and at the leading edge the specimen likewise experiences an enhanced compression" (Lawn and Marshall 1979:70). This created small cracks in the wake of the quartz fragment and forms partial herztian cones of percussion around the contact zone. The stresses in the immediate area are relaxed and a second ring or partial cone forms when the fragment has moved far enough to build up sufficient stress again. As the particle is dragged across the surface, a series of regularly spaced tracks result along the indentation path creating a linear groove. These linear grooves stop when the microchip either disintegrates or is displaced from the groove (Fedje 1979). This is similar to that seen on a grander scale in glacial movement, where ground rocks embedded in the glacier are forced along by the forward movement of the mass. Linear grooves are rasped out of the underlying rock and run parallel to the glacier's flow (Vial 1952).

Striae - These are linear features with smooth-bottoms caused when a small rounded fragment, often of silt or another extraneous particle is dragged across the surface of the tool or the polished area (Plate 17). Two types of striations were found on the quartz tools: a) narrow and deep; where the bottom of the trough is not visible at 200X and b) broad and shallow; where the bottom of the striation trough is clearly visible at 200X. Striae are also more regular in their overall configuration than linear grooves. It is likely that the rounded edges of the particles prevent them from getting "caught" in the bottoms of the troughs and creating the tracks seen in the bottoms of linear grooves.

Striae vary in their lengths depending upon the material being worked and the quantity of abrasive that is present (refer to individual sections for striae lengths). They can form very quickly, as soon as sixty seconds after use begins, but they are not necessarily cumulative over time. It has been noted (Del Bene 1979) that a softer material can produce striae on a harder one by transferring pieces from the surface of the softer material to the surface of the harder one, thereby appearing as if the striae have been smeared onto the harder substance. I did not find this feature on any of the experimental quartz tools even though there seems to be an "additive" polish on some of the flakes used for butchery (Plate 18). Certainly more research and continued study of the mechanics involved is necessary before more conclusive statements can be made.

Striae can occur by other means as well, for example, post-depositional soil movement can create random striations on the tool (see the section on "artificial effects"). They are distinguishable from use striae because they are random in their distribution and occur on the interior portion of the tool as well, far away from any working edge. When severe, post-depositional striae can obscure the microwear polish, making identification of tool use difficult and sometimes impossible.

Polish - The mechanics of polish formation remain an unresolved issue, and are thoroughly discussed in the literature (Anderson 1980; Bettison 1985; Del Bene 1979; Diamond 1979; Kamminga 1979; Mansur-Franchomme 1983; Moss 1983b; Rabinowicz 1968; Witthoft 1967). While the process of polish formation is beyond the scope of this research, it is worthwhile to present some of the current hypotheses concerning its development on stone tools.

A prevailing, and widely accepted, explanation for polish formation on metals comes from Rabinowicz (1968), and it appears that the same formation processes may be applicable to stone. Rabinowicz said that polish forms when molecules are removed by friction from the high points on the surface of the tool as it moves against the substrate, creating a smoothing of the surface. He detected a weight loss in the polished material which corresponded closely to the amount of material originally estimated to have been present on the high points prior to polishing. This provided an alternative hypothesis to the widely accepted theory of Beilby (1904), who suggested that polishing generated heat, which in turn caused melting and smearing of material from the high points on the surface of the metal to lower ones, bridging them with a smooth, thin skin, known as the "Beilby layer". While melting of the surface is not proven, polished metals were found to have "caps" of thin film on their surfaces which, when examined using electron diffraction, were found to be structurally different than the underlying metal .

I also found a similar situation on an archaeological piece of flint which I cross-sectioned and examined with the scanning electron microscope. This flint blade, from Kebara Cave in Israel, had an extremely well developed plant polish, or silica gloss, on its surface (Plates 19 and 20). When sectioned and examined at 1000X, an 'amorphous' layer, which was very different from the sectioned portion of unused flint viewed under the same conditions, was clearly visible on top of apparently unaltered flint (Plate 21). Research is in progress to determine if this layer is a deposition of silica onto the surface of the flint or whether another process, such as abrasion or melting has occurred. Others have observed a dissolution of the surface of the flint which, they say, is relative to the amount of heat generated by friction and the amount of moisture present in the material being worked (Anderson 1980 and Mansur-Franchomme 1983). On the other hand, Meeks, et al (1982), Diamond (1979), and Kamminga (1979), have suggested that some polish formation is the result of surface alteration caused by frictional mechanisms resulting in abrasion. Kamminga goes on to say that certain polishes, such as a soft plant polish, may be affected by the amount of moisture present. I too have found this to be the case.

Many researchers, when studying polish formation, concentrate their efforts on plant polish as it is the most easily recognizable of the polishes, is present on tools in both the Old and the New World and, when well developed, can be seen without the aid of a microscope.

Witthoft (1967) wrote that silica gloss was created when plant opals are deposited on the surface of the flint during use and then fused onto the surface by friction and heat. His theory does not encompass other types of polishes such as bone or wood but, the formation of other polish types has not been well studied to date. From this preliminary research on quartz, it appears that the polish left on a quartz tool after use on organic material is dependent upon moisture content, the presence/absence of abrasives and the hardness of the substrate. Through future research, we may discover that polish formation is, in fact, a combination of both deposition of material onto the surface of the tool and smoothing by attrition. Anderson (1980) suggested that it is the dissolution of the surface of the flint tool in conjunction with heat generated through friction and the amount of moisture in the material being worked that determines the degree of polish formation. Unger-Hamilton disagrees, stating that the plant phytoliths which Anderson says are melted on the surface of the flint are actually "relic grains of unreplaced skeletal calcite..."(1983:96) protruding out of the surface of the flint. I too have noted these 'intrusions' on at least one piece of unused flint examined at 200X using the SEM.

Specific polish types, their attributes and their formation are discussed in more detail in their individual chapters.

Edge Damage - This is the presence of small scars which have been removed from the edge of the tool during use. There are many ways, other than purposeful use, in which edge damage can occur. For example, small scars may be left on the edges of a flake during manufacture as the flake twists off the core. Damage may also occur as the flake hits the ground after being struck from the core or as a consequence of being stepped on or trampled. Post-depositional processes such as soil movement and pressure from overlying sediments can cause micro-chips to be removed from the flake or tool edges. Even curation techniques, if not carried out with a mind towards microwear studies, can create potential hazards for the use wear analyst by cleaning tools using wire brushes or storing many unwrapped tools togther in the same bag (Sussman 1985b). This is by no means an exhaustive list of "hazards" and one is referred to Keeley (1980) for a more thorough discussion of the problems.

Edge damage studies, or low-power studies as they are also called, have attracted much attention as a method for determining how a tool was used by the types of scars sustained on the working edge (Odell and Odell-Vereeken 1980; Semenov 1964; Tringham et al 1974 _inter alia_). It is the feeling of this researcher (see also Moss 1983a; Keeley 1974; Keeley and Newcomer 1977) that edge damage alone does not provide sufficient information to assess the function of a tool. In the absence of polish, it can determine (along with tool morphology and edge angle) in what manner a tool was used, such as sawing versus scraping, but can not be as specific as high-power microscopy in saying exactly _on what_ material the tool was used, e.g. bone versus antler. The study of archaeological material further compounds the problem because other processes such as tool resharpening or trampling alter and often mimic use-wear patterns (Flenniken and Haggarty 1979; Moss 1983a). This is not to say that high-power microwear studies are the definitive and final answer. The same problems which plague the low power analysts also create problems for microwear specialists by obscuring or removing use wear polish. High-power studies, however, are better able to transcend the broad categories of "hard versus soft" and potentially identify the specific materials on which the stone tools were used.

Ideally, one should employ a combination of the two methods when constructing a complete picture of the life history of a tool, as each method serves to enhance the other.

Scar Types

Keeley (1980) made a detailed analysis of the various minute scars he found on the edges of flint tools after use. It was simpler and just as useful to condense these categories when working with quartz.

Scars of different sizes were included in the same category if they had the same morphologcal features, as it is often difficult to focus on the edge of a quartz tool clearly and distinguish individual scar patterns due to surface reflectivity, edge irregularities, uneveness, and so on. I distinguished four gross scar categories along the edges of the experimental tools. These occurred in varying combinations and their presence depended upon the edge angle of the tool (is it relatively steep, or is it acute and will break easily?) as well as the activity the tool was used for and the material it was used on. The four scar categories are descibed below.

Scalar Scars - By definition (Crabtree 1972) these irregular and expanding scars resemble fish scales and are created when the force from use is applied roughly parallel to the flake's surface (Plate 5).

Step Fractures - "A scar that terminates abruptly in a right angle break...caused by a dissipation of force or the collapse of the flake" (Ibid:93). These scars were found most often after working hard, unyielding materials such as bone, wood or antler. Scraping hide produced step fractures along the tool edge aided, no doubt, by the log or wooden table the hide rested upon while being scraped (Plates 22 and 23).

Half Moon Fractures - These are caused by a bending, tensile force applied at right angles to the working edge, and are most often found on working edges having acute angles and thin cross sections. The scars resemble half moons or small "bites" taken out of the edge. They ceased to occur with regularity when the edge angle steepened as it stabilized during use.

Crushing - This was present with regularity on those tools used for graving and boring. Crushing destroyed the structure of the working edge by grinding it into minute particles which obscured the edge as well as any other scar patterns which might have been present. These tiny particles are not always removed by cleaning and may not be completely detached from the surface of the tool (Plate 24).

Tool Use Actions - These represent the manner in which the tools were used. I chose these actions in particular, because they were adaptable to a variety of materials, and allowed direct comparisons to be made between tools used in the same fashion on a variety of different materials (refer also to Keeley 1980).

Sawing - The working edge of the tool runs parallel to the direction of motion and is drawn bidirectionally back and forth. The sides of the tool are held close to right angles to the material being worked. The working edge on the tool was most effective when it was an acute, unretouched edge.

Cutting - consists of bi- or unidirectional strokes with the edge of the tool held parallel to the direction of use and at right angles to the material being cut. Cutting was the term used when the material worked is soft such as meat or grass, while sawing was used when the material worked was harder, as in the case of antler, wood or bone.

Scraping - The working edge of the tool is steeper than that used for sawing and cutting and is held close to right angles to the substrate as well as at right angles to the direction of use. The tool is pulled towards the user in a single motion.

Graving - the functional edge of a graver is referred to as the "bit" and is generally either of the two points at the tip of the tool. The tool is held parallel to the direction of use while being pushed away from the user. While this is a bidirectional action of pushing and pulling, there tends to be much less force exerted when the tool is pulled toward the user after the push-away stroke.

Boring - The tool is held perpendicular to the substrate and is rotated in a clockwise and counter-clockwise fashion. The force is exerted by the user down the long axis of the tool. The working edge of the tool is generally pointed for better perforation. Because of the crushing sustained on the tip of the borer, the direction of rotation of the borer during use could not be ascertained.

Wear Features

Attrition - This is the plucking off or rasping of the high points and edges of the tools leaving behind a rough, nubbly surface similar to sandpaper (Plates 28 and 29). It is extremely similar to abrasion but lacks striations and may be very restricted in its occurrence.

Abrasion - A process of erosion that involves rubbing the surface of a tool against another substance (such as wood, meat, bone, etc), creating friction and generating heat. This results in the removal of surface material, leaving behind a roughened surface, somewhat similar to attrition, but with accompanying striations and linear grooves.

"Artificial Effects" - There are many ways in which pseudo use wear may be produced on a flake without having been used. These effects can seriously interfere with microwear identification and cause mistakes. It is important for these traces to be identified and recognized as belonging to other processes which have nothing to do with utilization. Below are some of the primary ways in which tools may be altered by either natural or "human" agencies (Keeley 1980 has a good overview; see also Yamada and Shimura 1984).

Hammerstone Smears - These are often localized near the platform of the flake and occur when a flake is removed from the core, or while adjusting the angle of the platform before striking it to remove the flake. These "smears" are very thin, multiple parallel striae and appear as bright patches on flint. On quartz, hammerstone smnears may also be accompanied by crushing and spots of extremely restricted abrasion. Generally, they are easily identified and do not interfere with use wear identification.

Hard Hammer Retouch - This is the systematic removal of small flakes along the edge of a tool (usually the edge to be used) which serve to regularize edge morphology and modify the edge angle as well. Retouch may mimic utilization if the scars are very small and continuous down the edge of the tool but, this is not often the case. It is also possible to find hammerstone smears along the retouched edge when a hard hammer has been used for retouching.

Twisting off the Core - If the core is held so that the flake cannot fall away from the core as it is struck, (being held in place by the fingers), it may twist against the core and leave abrasion tracks and small scars. This "spontaneous retouch" (Newcomer 1976), also occurs when a flake strikes the ground during manufacture. While abrasion tracks are easy to spot on flint and relatively so on quartz, the edge damage produced may be harder to recognize. Tringham (1974) feels that distinctions can be made between scars from retouch and those from use on the basis of patterning and size.

Trampling - Being stepped on by people or animals after discard can create edge damage on artefacts which closely resembles edge damage from utilization or retouch (Flenniken and Haggarty 1979).

Soil Movement - This process occurs post-depositionally and may leave localized traces of abrasion and/or striations on the tool as well as removing small, randomly situated scars along the tool edges. Abrasion may look like bright, slightly greasy areas along the edges and high points of the microtopography and can obscure areas of weakly developed polish. It may also take the form of crushing and/or attrition on quartz tools.

Patinations - A chemical alteration of the surface of the stone which obliterates traces of microwear polish. Patination comes from moisture leaching from inside the stone after prolonged exposure to the air, or from chemicals in the soil after burial. It may also be caused by organic material adhering to the surface of the stone. This can be a problem when dealing with flint tools but, quartz is chemically stable and suffers instead from abrasion, which can easily obscure polish traces. Some patination and chemical encrustations on flint can be removed by prolonged or heated chemical soaks (see cleaning procedure).

"Desert Varnish" - Long exposure to the elements and wind, in particular, causes a chemical alteration of the tool's surface, creating a patina. I have seen this phenomena on flint tools but not on quartz. Severe abrasion, however, does occur in the desert from the particles of sand carried by the wind continually blasting away at the surface of the exposed rock. This process can drastically affect the surface of the quartz.

"Rolling" - The continual movement of water over stone along with the sediments carried by the moving water, causes ridges and edges of stone tools to beome blunt and rounded, acquiring a nice sheen (Schackley 1974). The tools may also bang against rocks and boulders as they tumble along in the water, removing new flakes from their edges.

"Bag Damage" - When stone tools are curated after excavation, they are often stored unwrapped, together in bags. The sheer weight of the stones can cause small chips to be removed from their edges. While these are randomly located and unrelated to areas of use, the jostling of unwrapped tools against one another in bags or in trays, abrades the ridges and edges of the tools so that traces of use may be permanently obscured (Keeley 1980). Bag damage may mimic edge damage from utilization as well.

Statistics

Mean and standard deviation were run on the raw counts of the attributes (numbers of pits, number of striae, etc). To get these counts, I chose a total of five areas on each tool and counted all the attributes I was able to find in an area measuring one millimeter across. This corresponded to the size of the area visible in the eyepieces of the light microscope at 200X. Unfortunately, these areas could not be randomly chosen, due to the irregularity of the quartz surface and extreme variability in reflectivity and surface texture. I was forced to choose those places where: a) I could focus equally well on the entire area and b) I knew that that area on the tool had contacted the material worked. As a control, prior to the actual counting, I examined fifty areas on various tools I knew had **not** been used, to make certain I could find no attributes I assumed to be caused soley by use.

These statistics were calculated to determine if a relationship existed between the length of time the tool was used, the material it was used on, and the number of attributes on the surface of a used quartz tool. The original hope was that this might provide additional information for identification purposes. Many significant relationships emerged but, while interesting, they are useless for identification of tool use because:

1) Quartz, as a raw material, is too variable in its surface texture to provide consistent results regarding the numbers of attributes present. In other words, one would not be able to tell that a quartz tool was used to scrape a dry hide, for instance, by the number of pits or linear grooves present on the used surface.

2) Stone tools were used by people not machines, therefore, the amount of pressure exerted on a tool during use is not constant. This variance cannot be accurately measured to support these statistics, and the goal in researching this topic was to understand what happens to the surface of quartz when used by individuals.

Therefore, the results of these tests and the regression plots are inserted because of the relationships they illustrate and the information they convey but, unfortunately, microwear analysts will not be able to look at the average number of pits a quartz tool has and determine on that basis alone what the tool was used for.

Plate 1. Unused Crystal Quartz Surface
Light Microscope 200X

Plate 2. Unused Crystal Quartz Surface
Light Microscope 200X

Plate 3. Unused Crystal Quartz Surface
Light Microscope 200X

Plate 4. Unused Crystal Quartz Surface
Light Microscope 200X

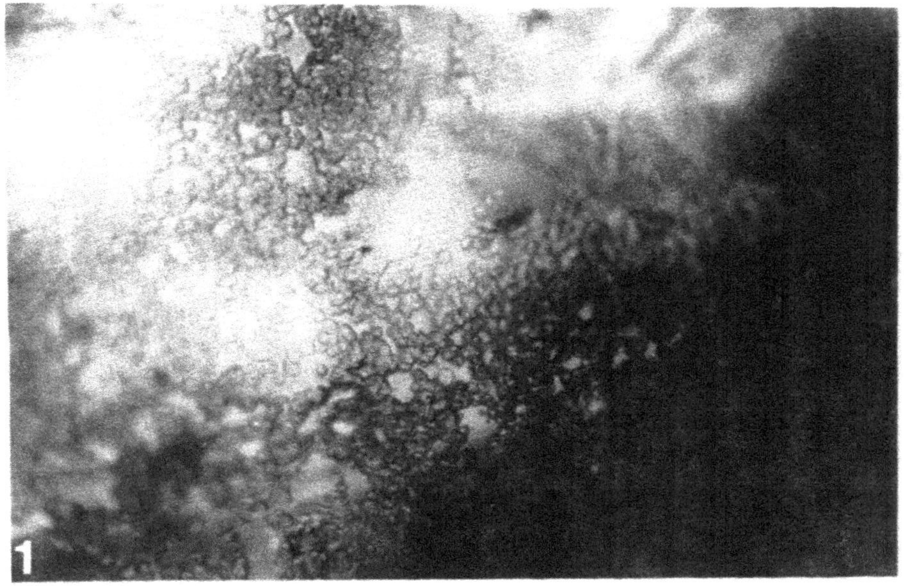

1

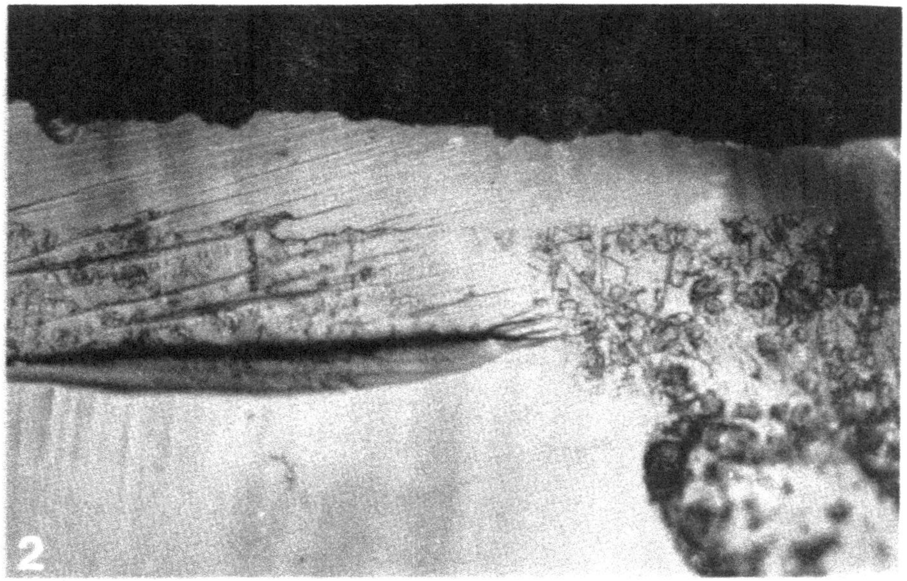

2

3

4

Plate 5. Scalar Scars on Unused Crystal Quartz Edge
Light Microscope 200X

Plate 6. Crystal Quartz Edge used for Cutting Grass, Tilden Park
Light Microscope 200X

Plate 7. Scanning Electron Microscope Photo
same area as Plate 6. 200X

Plate 8. SEM Photograph. Enlargement of Plate 7 1000X

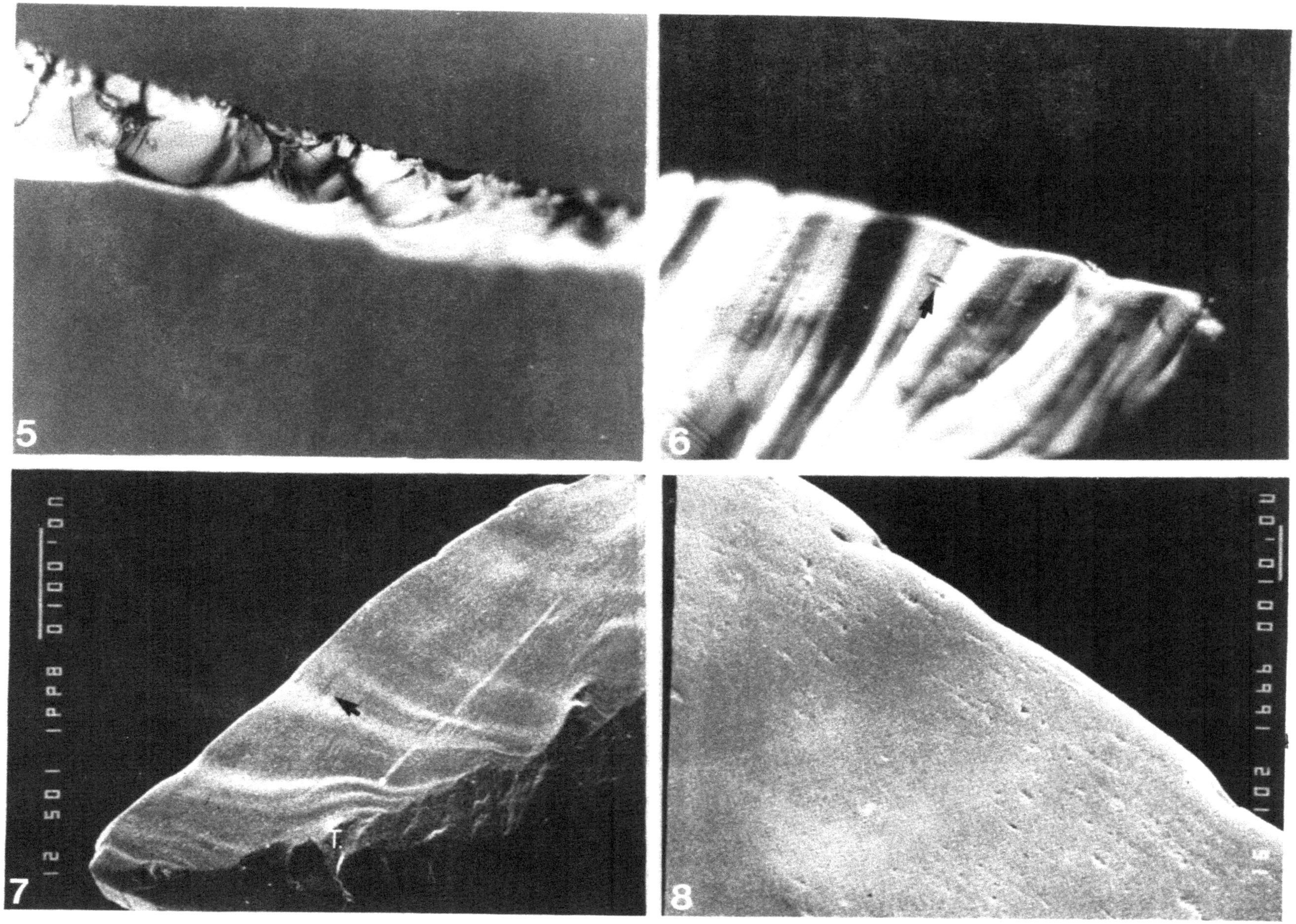
5
6
7
8

Plate 9. SEM Photo of Used Quartz Edge with Comet-tailed Pits 30 degree tilt 500X

Plate 10. SEM Photo of Quartz Edge after Scraping Fresh Hide with Volcanic Ash. 0 degree tilt 200X Arrow points to same feature in Plates 10–13

Plate 11. SEM Photo. Same as Plate 10 but tilted 20 degrees Note change in features due to tilt 200X

Plate 12. SEM Photo. Same as Plates 10 & 11 but tilted 30 degrees. 200X

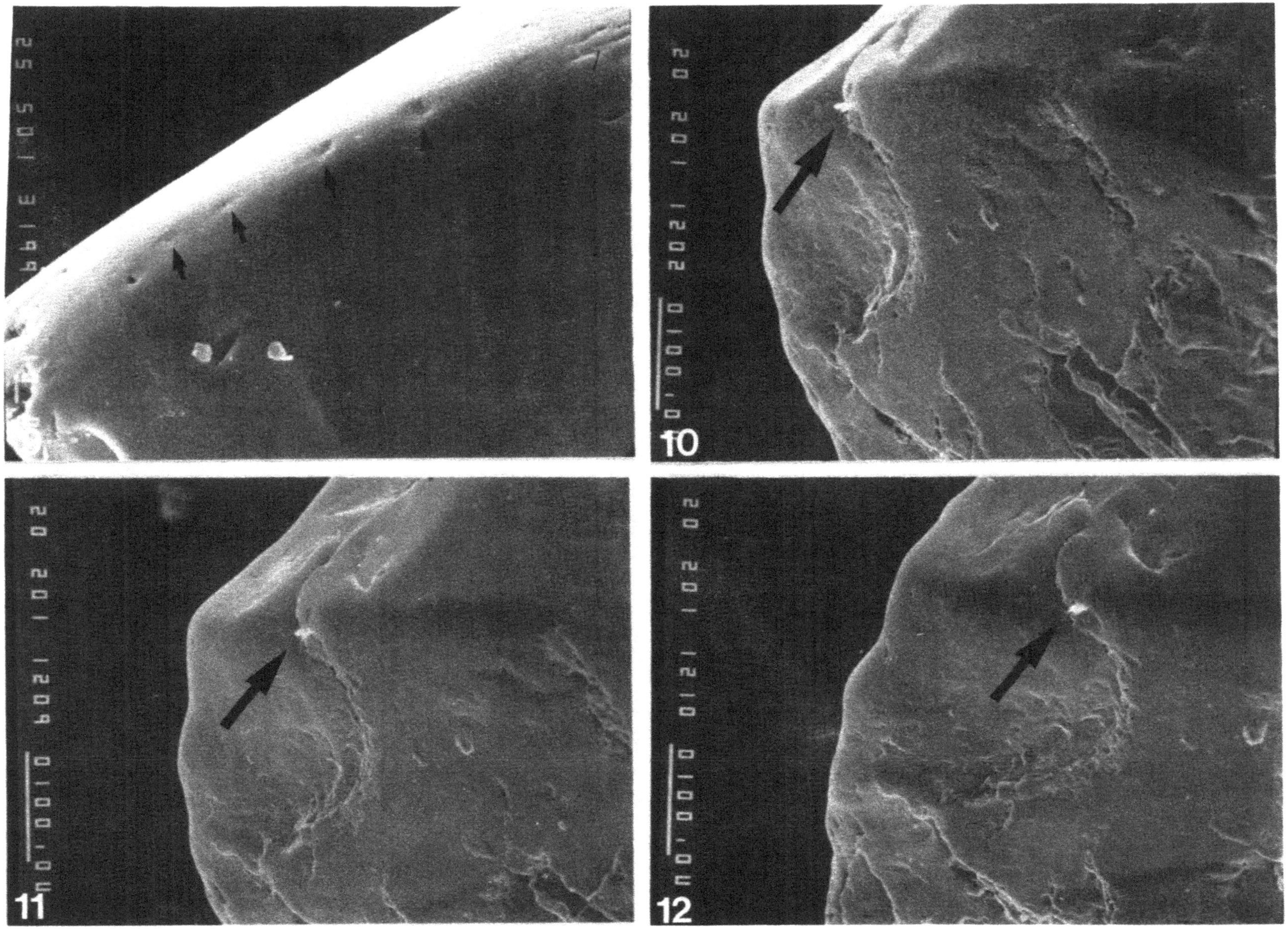
10
11
12

Plate 13. SEM Photo of Quartz Edge after Scraping Fresh Hide with Volcanic Ash. 45 degree tilt 200X
Arrow points to same feature in Plates 10-13

Plate 14. SEM Photo. Used Crystal Quartz Edge showing "roches mountonees" 500X

Plate 15. Used Crystal Quartz Edge showing Comet-tailed Pits Light Microscope 200X

Plate 16. SEM Photo. Used Quartz Edge with Linear Grooves 200X

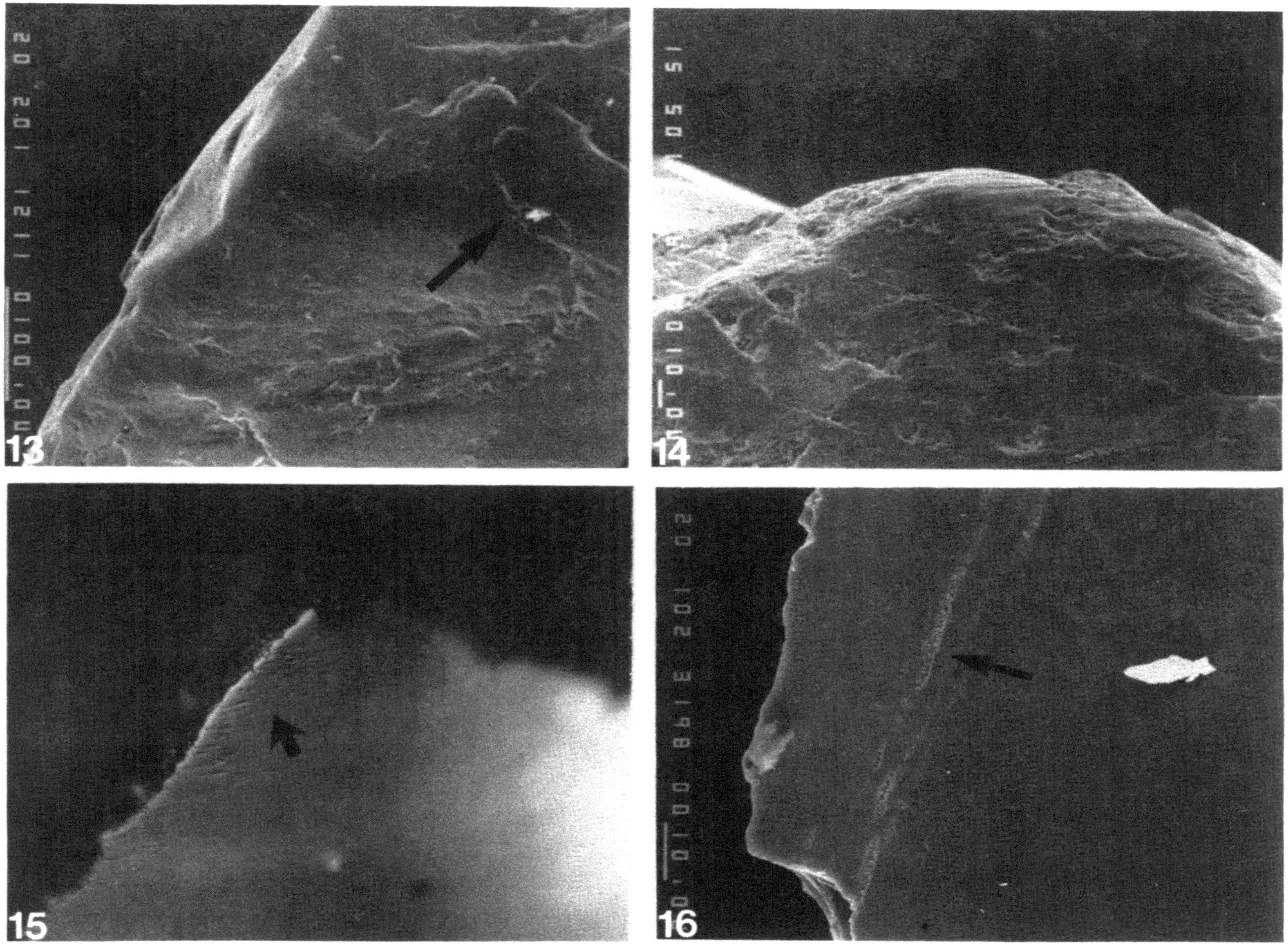
13
14
15
16

Plate 17. SEM Photo of Used Quartz Surface with Striation 3900X

Plate 18. Crystal Quartz Edge Used for Cutting Meat 500X Note ?additive material. Light Microscope

Plate 19. SEM photo. Cross-sectioned Flint Blade with Silica Layer ontop of apparently unaltered flint. 1000X

Plate 20. SEM Photo. Enlargement of Plate 19. 10,000X

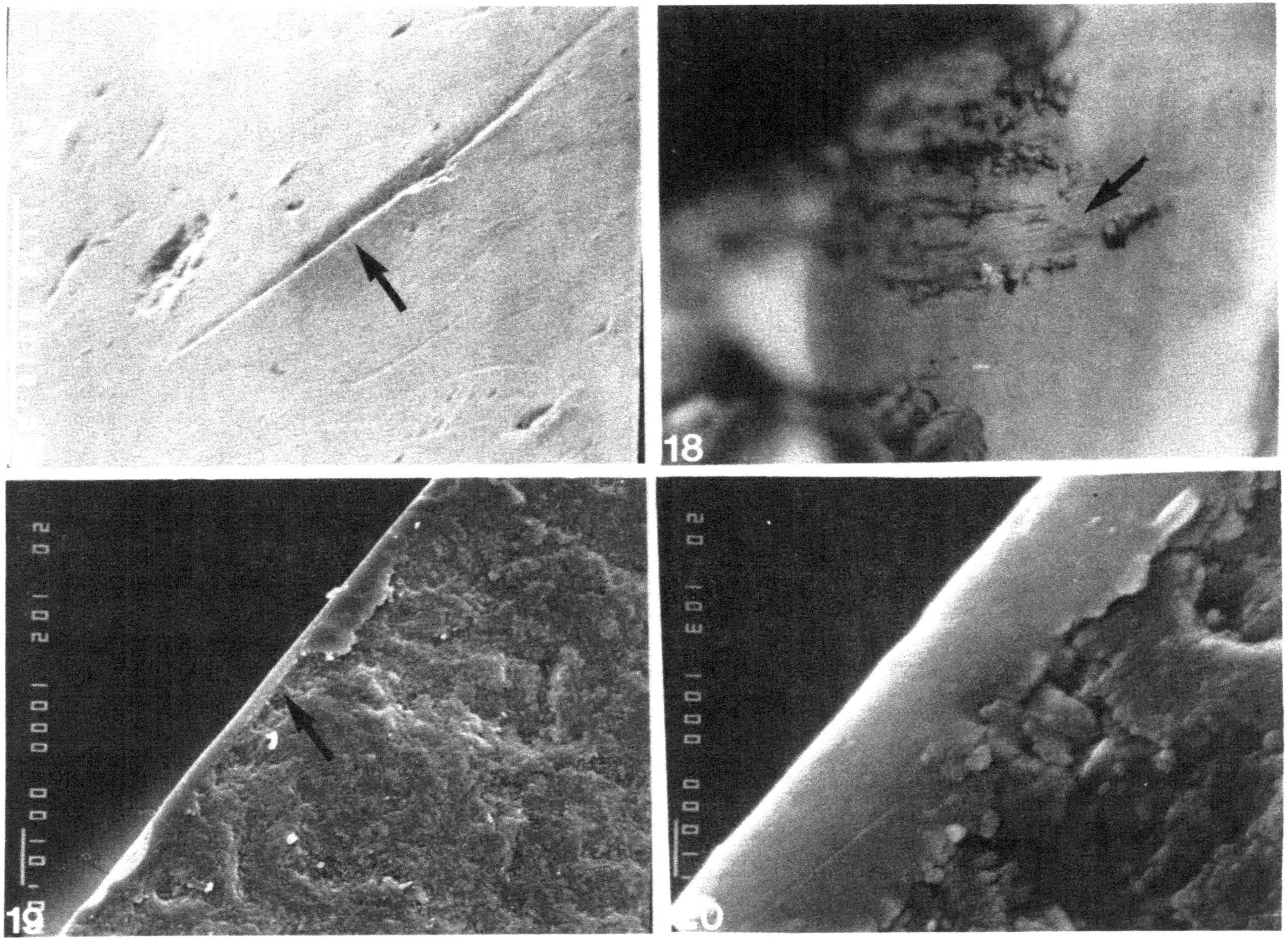
18
19
20

Plate 21. SEM Photo. Cross-section of Unused Flint Blade 1000X

Plate 22. SEM Photo. Step Fractures on Used Quartz Edge 100X

Plate 23. SEM photo. Step Fractures on Used Quartz Edge 1000x

Plate 24. SEM Photo. Crushing on Used Quartz Edge 200X

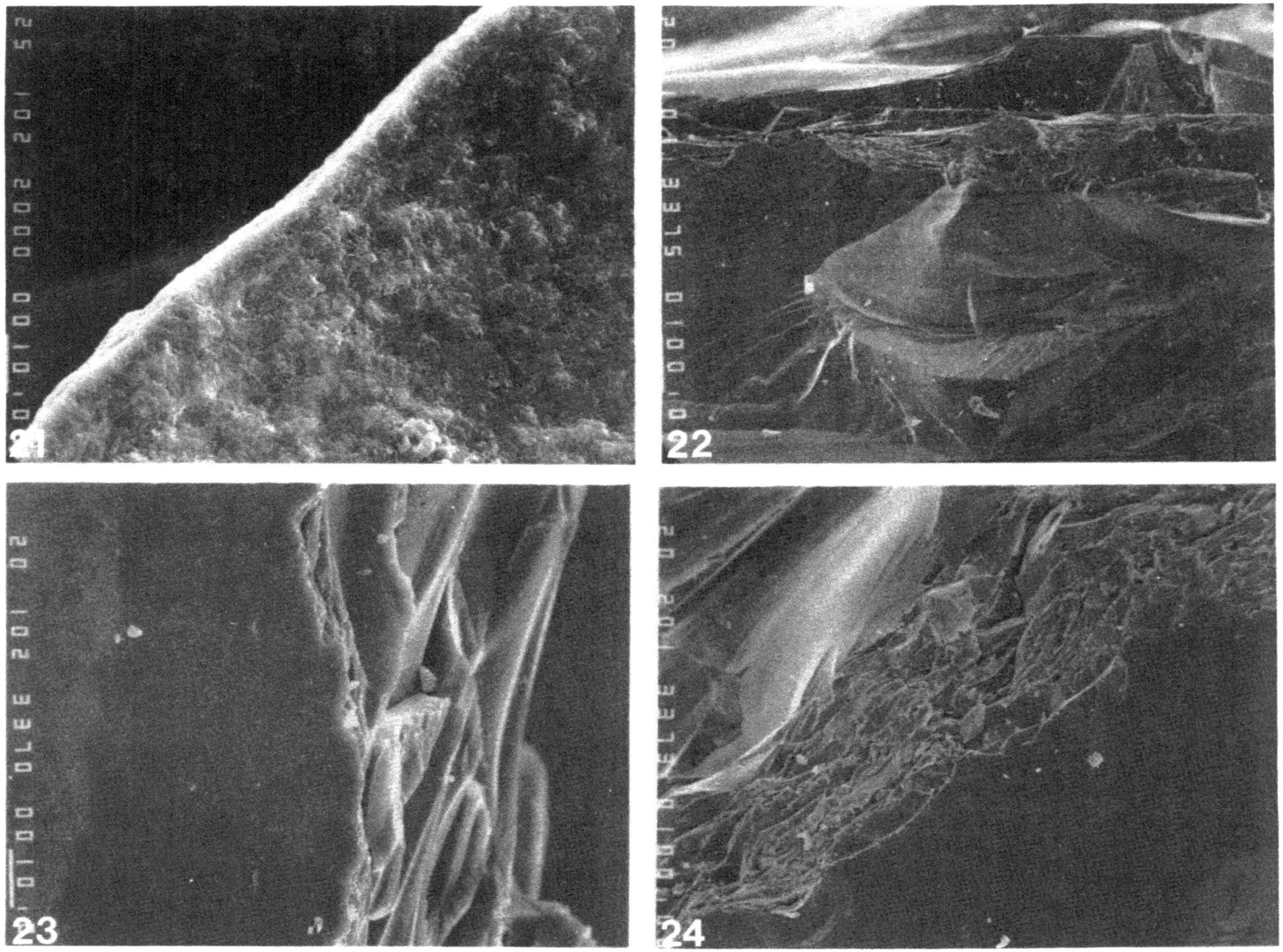
21
22
23
24

Chapter 3

Hide Working

Total Number of Flakes: 25

Total Number of Attribute Counts (5 per piece): 125

Number of Flakes Scraping Fresh Hide: 5

Number of Flakes Scraping Sun-dried Hide: 15

Number of Flakes Scraping Tanned Hide: 5

Scars - Scars were present only on the side of the flake opposite that which came in contact with the hide. Both scalar and half moon scars were visible after one minute of use on all three types of hide (wet, sun-dried and tanned), and became more numerous as use time increased. Step fractures were evident after ten minutes of use on both the tanned and sun-dried hide tools but, were never seen on those quartz tools used to scrape wet hide. The side of the flake scraping the hide (usually the ventral), sustained little or no edge damage.

Pits - Scraping sun-dried hide produced many more pits, on average, than either tanned hide or fresh (wet) hide (Table 1). This could be due to the comparatively hard and stiff condition of the sun-dried hide causing more abrasion and friction and, hence, more pitting on the surface of the quartz than either fresh or tanned hide.

Few pits were produced on the quartz tools used to scrape fresh hide (Plate 25). As fresh hide is moist and pliable, perhaps it was not possible to exert sufficient pressure on the stone tool to cause deformation of the tool's surface or "surface fatigue", thereby causing pits to form. Friction generates heat which, if present, would presumably cause micro-fragments to spall off the surface. Although a novice, I quickly realized that scraping wet hide was not a productive task if one's goal was to remove the inner layer of membrane and subcutaneous fat. The result was, as another researcher aptly phrased it, akin to scraping rubber cement that had not yet dried; not much was removed. Scraping fresh hide was most productive after the hide had been pegged out in the shade for a few hours. It was still pliable, but not as wet. When left in direct sun to dry, the hide became extremely stiff and difficult to work. At that point, it would have been easier to rub grease or fat back into it to render it maleable.

It has been documented (Speck 1937: 345-347) that Eskimos process seal skin while still "wet and raw" to preserve its waterproof quality. They originally used seal scapulae for scrapers, not stone. Other peoples also employed bone tools; the Kwando peoples, mixed farmers living in the southeastern part of Zambia used maxilla, parts of crania and pieces of mandible to scrape hides (J.D. Clark pers. comm.). Clark (1958:144) mentions ethnographic reports from the late nineteenth century documenting Bushmen using stone endscrapers for skinning game, "...though there is no reason to suppose that this was the only use to which these tools were put." Clark also mentions Dr. Kannemeyer, who described the skinning knives used by the eastern Free State Bushman in 1890. These were Smithfield end scrapers, "...wrought flakes used for skinning game...They exactly resemble certain scrapers in use amongst the Esquimo, but were never affixed to a handle.." (Clark 1959:232). Van Riet Lowe's description of the Old Bushman from the Winberg district of the Orange Free State is also pertinent,

> "...he told of his father's use of stone for the manufacture of scrapers for paring down and shaping wooden clubs, bows and arrows, for cleaning skins, preparing karosses and taking the meat off bones..."

(Ibid:232).

The Ukwe (San) are also known to process pegged hides by scraping them with a mixture of brains using an upper grindstone, while the pastoralists at Errer Springs in Ethiopia used bricks for scraping hides. Small edges of the brick would crumble, creating the necessary grit needed for efficient scraping (J.D. Clark pers. comm).

Hide scraping was the first set of experiments I performed in my research. I was in the state of Washington where Mt. St. Helen's had recently erupted, leaving behind large amounts of volcanic ash. After fruitlessly scraping fresh hide in my first experiment, I scooped up a handful of the ash and, in frustration, dumped it on the hide in an attempt to increase scraping efficiency. As mentioned above, there is evidence for the introduction of extraneous material into hideworking to aid in processing (see also Brink 1978). Adding volcanic ash to a fresh hide is not likely to have happened archaeologically but, in areas such as East Africa where volcanic activity was prevalent more than a million years ago, ash certainly was present - sometimes in copious quantities. In Washington, for example, there was still a good bit of ash on the ground more than a year after the eruption. Hence, there is no reason to doubt that ash permeated most things. The results of this experiment were the first to document definitively the presence of use-wear on quartz. (They are discussed in more detail later in this section).

Scraping sun-dried hide produced a greater number of pits overall than scraping either wet or tanned hide; these pits increased significantly in number as the time of use increased ($p<.01$ Fig. 4). This was documented microscopically as well, prior to counting the number of pits (see Plate 26). Commercially tanned hide was scraped using quartz tools for comparative purposes, to see if there were differences in wear-attributes when compared with quartz scrapers used on sun-dried hide. I assumed there would be little, if any, difference between the two but, as mentioned, scraping tanned hide produced many fewer pits overall (Plate 27). Perhaps tanned hide, being softer and more pliable than sun-dried hide, created less friction, and therefore less heat when scraped resulting in less pitting. This made sense until statistics showed that scraping fresh hide produced more than twice the mean number of pits than tanned hide! (Table 1) It is possible that a greater number of "incipient" weak spots present in those flakes used to scrape fresh hide caused an increase in pitting although this seems unlikely. One difference between the two experiments is that the flakes used to scrape fresh hide were all made from vein quartz, while the set of flakes used on the commercially tanned hide were of crystal quartz. While it is conceivable that vein quartz sustains more pits under certain conditions, this was not a consistent finding when both vein and crystal quartz were used on the other materials. The chemicals used in the modern tanning process of leather: alum, brine solution, aluminum sulfate, sulphuric or oxalic acid, may, in some way, affect polish formation as well.

Numbers of Pits

	N	Mean	S.D.	Range
All Hide Working	120	39.06	54.85	0-30
Sun-dried	75	51.02	65.14	0-300
Tanned	25	11.68	14.33	0-60
Fresh	20	24.75	20.62	0-65
Fresh w. ash	5	699.8	241.74	300-900

N refers to the number of observations (five per piece).
* All hide working minus the piece used with volcanic ash.

Table 1

Striae - Scraping all types of hide, excluding fresh hide, resulted in striae forming on the ventral surfaces of the quartz scrapers to varying degrees. There was certainly dirt and grit around while the fresh hide was being scraped but, it was too wet and pliable for striae to occur. Mansur-Franchomme (1983) has suggested that very few striae form because the amount of moisture present when wet hide is scraped causes the surface of the tool (flint in her case) to "melt", becoming gel-like. Hence, as a striation forms, it immediately fills in. I was unable to replicate her findings on quartz; both the light microscope and the SEM show no evidence of surface alteration of the quartz scrapers after use. If, indeed, melting of the surface is occurring, there was no visible infilling or covering of other surface features and blemishes on the areas of the quartz which contacted the wet hide during use. I suggest that, if that is the case, there is not enough friction and, hence, enough heat generated being generated to raise the surface temperature enough for melting to occur. More likely, striations are not found on the surface of quartz tools used to scrape wet hide because the material is too wet for grit to become incorporated as an abrasive agent - rather it "rolls" along, cushioned by the moisture and suppleness of the hide. Once the hide begins to dry, however, there must be a threshold at which the abrasives will find purchase and create visible striae.

Scraping tanned hide, on the other hand, (disregarding the flake used with volcanic ash), produced the longest striae of all three types of hide, in addition to severe edge attrition. This was presumably due to a lack of moisture in the hide (see Table 2).

The numbers of striae produced while scraping tanned hide increase significantly over time (p<.001 Fig. 5). This was not the case, however, for sun-dried hide, where the greatest number of striae occurred after five minutes and again after an hour of use (see Appendix 3).

The lengths of striae produced while scraping tanned hide also increased significantly over time (p<.001, Fig. 6). All striae were measured at 200X with a micron scale under the light microscope. This was the only significant statistical relationship present when correlations were run on amount of use and striae lengths. Mean striation length for all hide working is shown in Table 2. Scraping fresh hide with volcanic ash produced the longest striae but, it should be noted that while eleven striae were counted, only two were measured. This is not enough to be statistically significant and is mentioned for interest only.

The range of striae lengths produced by scraping tanned hide is four times greater than those found after scraping sun-dried hide. Why this is the case remains unclear. One might expect striae to be longer on those tools used on sun-dried hide as wind-borne particles could be introduced as the hide was drying.

Mean Striae Lengths for Hideworking

	N	Mean	S.D.	Range
All Hide Working	181	8.71	9.89	0-100
Sun-dried	98	7.26	5.13	0-25
Tanned	83	10.43	13.30	0-100
Fresh	0			
Fresh with ash	2(11)	12.0	8.0	4-20

N refers to the number of striae measured.
* not included in total of all hide working

Table 2

Numbers of Striae

	N	Mean	S.D.	Range
All Hide Working	120	3.39	8.16	0-60
Sun-dried	75	1.53	2.94	0-12
Tanned	25	11.68	14.33	0-60
Fresh	0			
Fresh w.ash	5	2.2	.74	1-3

N refers to the number of attribute observations (five per piece).
* not included in total of all hide working.

Table 3

The majority of the striae produced by scraping hide are broad and shallow, but a few narrow and deep ones are present as well. Both types of striae were found after scraping sun-dried and tanned hide with quartz tools. Striae are found on flint tools used for scraping hide (Keeley 1980; Mansur-Franchomme 1983) but, they are somewhat different, being narrow and deep with rough-bottoms.

The direction of use of the tool was inferred by the orientation of striae to the working edge; most were perpendicular, while a few were situated at acute angles to the edge (<50 degrees). These often criss-cross one another due, undoubtedly, to the experimenter altering her grip on the tool during use.

Linear Grooves - These became visible on the quartz surface after one minute of use, and were present continually thereafter only on those tools used for scraping sun-dried or tanned hide. Linear grooves were oriented at right angles to the working edge of the tool along with the striae. Overall, linear grooves were more numerous after scraping sun-dried hide when compared with those seen after scraping tanned hide (Table 4). When the numbers of linear grooves from all hide working were combined, a positive correlation between tool use, time and linear grooves was found (Fig. 3). The same is true when tools used for scraping sun-dried hide were examined separately. It is interesting, that after scraping sun-dried hide for only one minute, the mean number of linear grooves observed was relatively high (9.8 ± 11.6) compared not only with fresh hide and tanned hide (3.2 ± 2.1 and 0 respectively), but with the other time periods up to one hour of use (refer to Appendix 3). Because of the futility of the activity, fresh hide was not scraped for more than twenty minutes.

Numbers of Linear Grooves

	N	Mean	S.D.	Range
All Hide Working	120	4.65	7.22	0-38
Sun-dried	75	6.48	8.42	0-38
Tanned	25	2.12	2.84	0-12
Fresh	20	1	1.73	0-6
Fresh w.ash	5	1	0.4	0-1

N refers to numbers of attribute counts (5 per piece).
* not included in total of all hide working.

Table 4

Ridge Attrition- Attrition of the ridges appeared after scraping sun-dried or tanned hide for five minutes and was noted on the remaining hide-working tools after ten minutes of use. While present, attrition remained a minor attribute, even on those quartz scrapers used for sixty minutes. As discussed in the section on wood working, the presence and severity of attrition corresponds to high points on the microtopography of the flake coming into contact with the substrate during use, and its absence should not mistakenly be used as an indicator of either "lack" of use, or use on a soft material.

Edge Attrition - Edge attrition was very slight after one minute of scraping sun-dried or tanned hide, and became noticeable on fresh hide after ten minutes of scraping. It was present on all of the tools after twenty minutes, regardless of the type of hide they scraped, but attrition of the edge was more severe on those tools used for scraping sun-dried or tanned hide (Plate 28).

Edge Rounding - Rounding of the worked edge was a consequence of edge attrition after scraping sun-dried or tanned hide for ten minutes. This feature was never seen on any quartz tool used to scrape wet hide (Plate 29).

Both rounding and attrition of the edges of flint tools depend upon the presence of abrasive agents on the hide (Brink 1978; Mansur-Franchomme 1983) so that dirt, grit, and micro-chips of quartz probably also cause rounding of the edge of quartz tools used for the same purpose.

Surface Texture - The only change in surface texture noted was roughening, seen on a quartz tool used to scrape sun-dried or tanned hide for five minutes. When used longer than five minutes, attrition of the edge became more prominent and invasive. The surfaces of all the tools were not consistently affected in this manner.

Polish - The experimental flint flakes sustained "typical" hide polishes after use (Keeley 1980). Those flakes used to scrape fresh hide corresponded well to Keeley's description for fresh hide polish; being somewhat bright and greasy (Ibid:49, Plate 30). The fresh hide polish was better developed and visible earlier than the polish on flint scrapers used to work either tanned or sun-dried hide (Plate 31). Fresh hide polish on flint resembles a meat polish, although rougher and less invasive. Edge damage on the scraper was restricted, for the most part, to the side of the flake opposite the worked edge. Scraping sun-dried and tanned hide with Brandon flint produced polishes that were dull and pitted, having matt surface textures. Although the tanned hide polish was also slightly greasy in lustre, only two quartz flakes used to scrape a fresh hide showed any evidence of polish; slightly dull and rough, lacking the greasy luster which is characteristic of the same polish on flint. It may be that the highly reflective quality of the quartz obscures the greasy luster.

The scrapers used on sun-dried hide and tanned hide, both sustained dull, greasy, rough-textured polishes, while only the quartz flake used on sun-dried hide for ten minutes had a bright, rough line visible down the worked edge at 100X. Scraping tanned hide with quartz did not result in any discernable polish.

In a 1983 report, Mansur-Franchomme suggests that polish formation on flint is enhanced by the amount of moisture and abrasives present in the material being worked. (see polish definition in Chapter 2). Essentially, the wetter the substance to be worked, the more quickly the polish will develop and the brighter it will be. This was also suggested by Patty Anderson in an earlier work (1980), where she found dissolution of the working edges of flint to occur during use, forming an amorphous layer of polish. The formation of this amorphous layer is dependent upon various factors, the most important of them being the amount of moisture present in the worked material. The next most important is the type, (and probably the amount) of abrasives present.

One of the investigators (M-F), scraped dry hide using flint and added water as a lubricant. After use, she described a polish which "showed the dull pitted aspect typical of hide-working tools, but it was slightly more brilliant than the polish formed when the hide was worked without water....the polish was very well developed and looked rough and pitted" (Mansur-Franchomme 1983:226). In an attempt to re-create her results, I re-wet a sun-dried hide by soaking it in water for twelve hours, and scraped it for twenty minutes using flint and three quartz flakes; one for one minute, one for sixty minutes and one flake sequentially for one, five, ten, twenty and sixty minutes. I cleaned and examined the flake used sequentially after each time sequence, using both the light microscope as well as the scanning electron microscope. The flint scraper sustained a few moderately bright, slightly greasy spots while the quartz flakes (minus the flake used for one minute which had no discernable evidence of use), sustained some pitting, a few striations and edge damage in the form of scalar and step scars (Plate 32). There was no rounding, alteration in surface texture, or polish visible on any of the quartz flakes studied using the light microscope, nor were there any changes on the flake used sequentially when examined with the scanning electron microscope after each time segment. Mansur-Franchomme's article does not mention the amount of time each flake was used but, I chose to use the flint scraper for twenty minutes as experience indicated that was a sufficient amount of time for a good, easily recognizable hide polish to form on a piece of flint. From her conclusions, Mansur-Franchomme states,

> "When abrasive agents are not present and all the other variables are kept constant, the amount of dissolution of the tool edge and, thus, the development of the micropolish, are a function of the amount of water present in the material worked," (Ibid:228).

From these findings it is unclear why the flint scraper did not develop a "brilliant" polish unless the soaked hide somehow lacked sufficient moisture.

Scraping fresh hide with volcanic ash added as an abrasive - As mentioned previously, I used a flake of vein quartz to scrape fresh bull hide for ten minutes and added volcanic ash as an abrasive agent to facilitate the removal of the inner layer of fat from the hide. While this method is unlikely to have been a regular practice in prehistory, the process did produce startling results (Plates 33 &34). As shown in Tables 1, 3 & 4, a great number of pits was produced on the quartz scraper, although very few striae or linear grooves were found. The polish was similar to a typical dry hide polish, being dull and having a matt texture (Keeley 1980:49; Mansur-Franchomme 1983:226). The used quartz edge also was markedly rounded and pitted. I found no potlid fractures on the quartz surface in contrast to Keeley who saw these on his flint hide-working tools (Ibid:49-50). Microprobe analysis of the polished surface was unsuccessful in revealing any residues, either volcanic or organic.

Summary - Consequently, the two main or prominent features of hide polish on quartz are: severe edge attrition and surface abrasion. Orientation of the striae and linear grooves aid in identifying direction of use. I found no polish to be present, unfortunately, which could aid in differentiating a hide polish from other polishes whose attributes might overlap during certain stages in their development, such as a wood polish.

Regression: Linear Grooves on Time for Hide Working

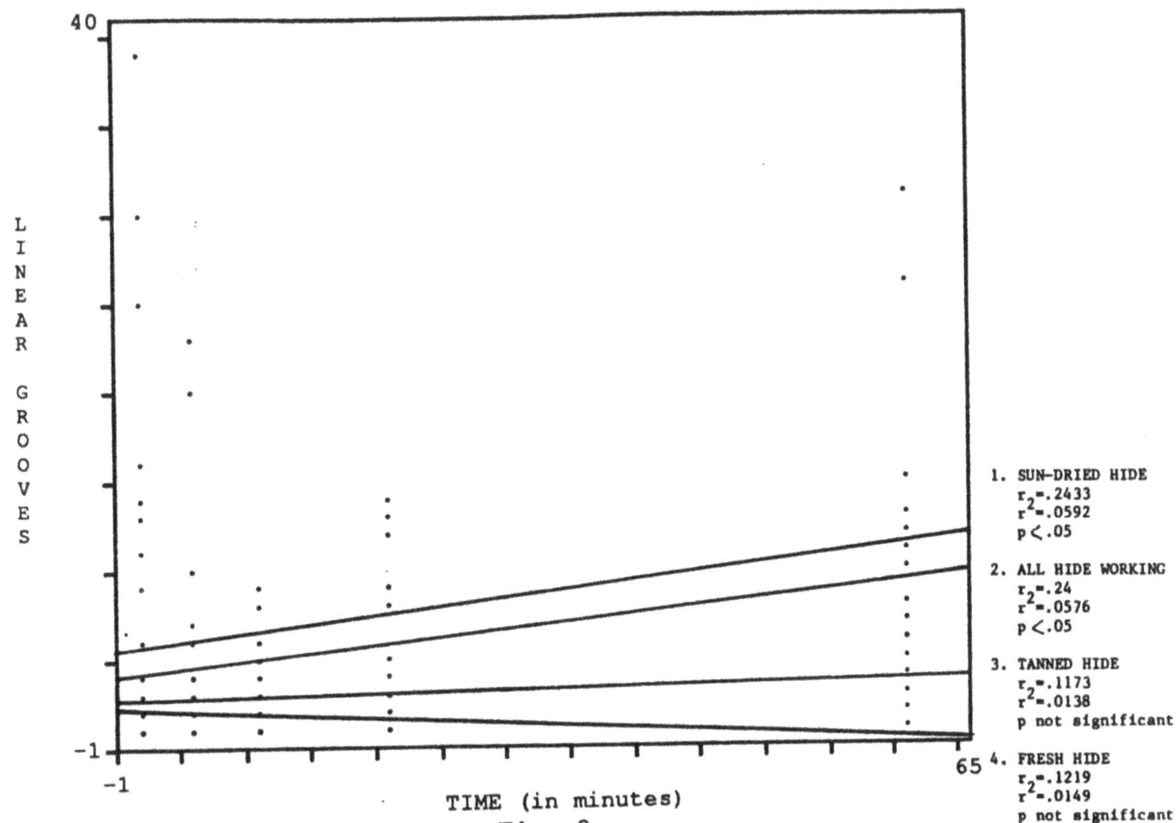

Fig. 3

Regression: Pit Formation on Time for All Types of Hide

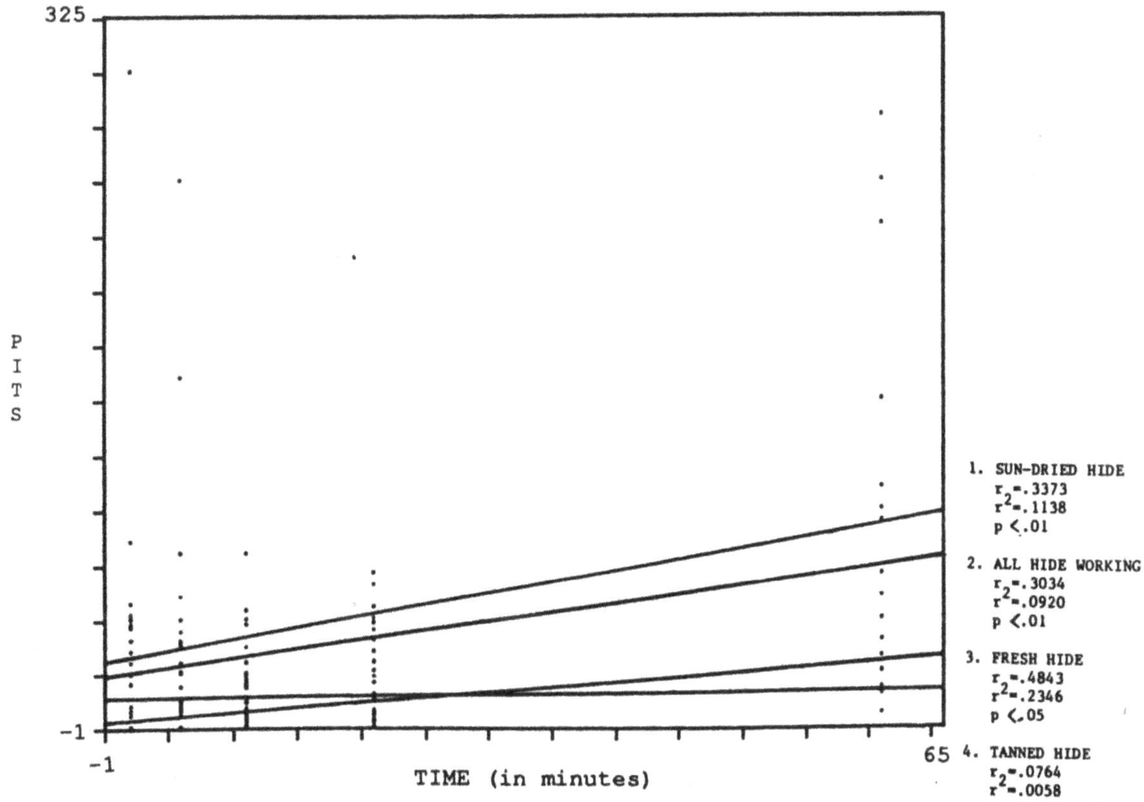

Fig. 4

Regression: Striae Formation on Time for All Types of Hide

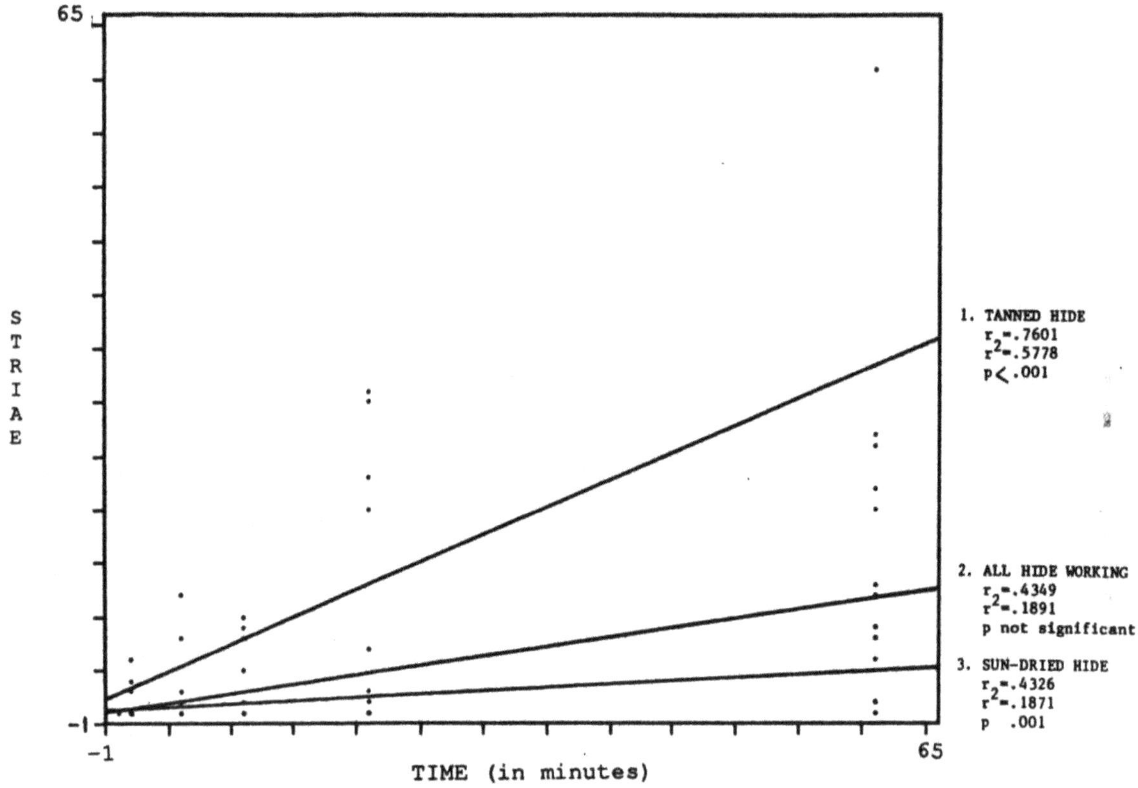

Fig. 5

Regression: Striae Length on Time for All Hide Working

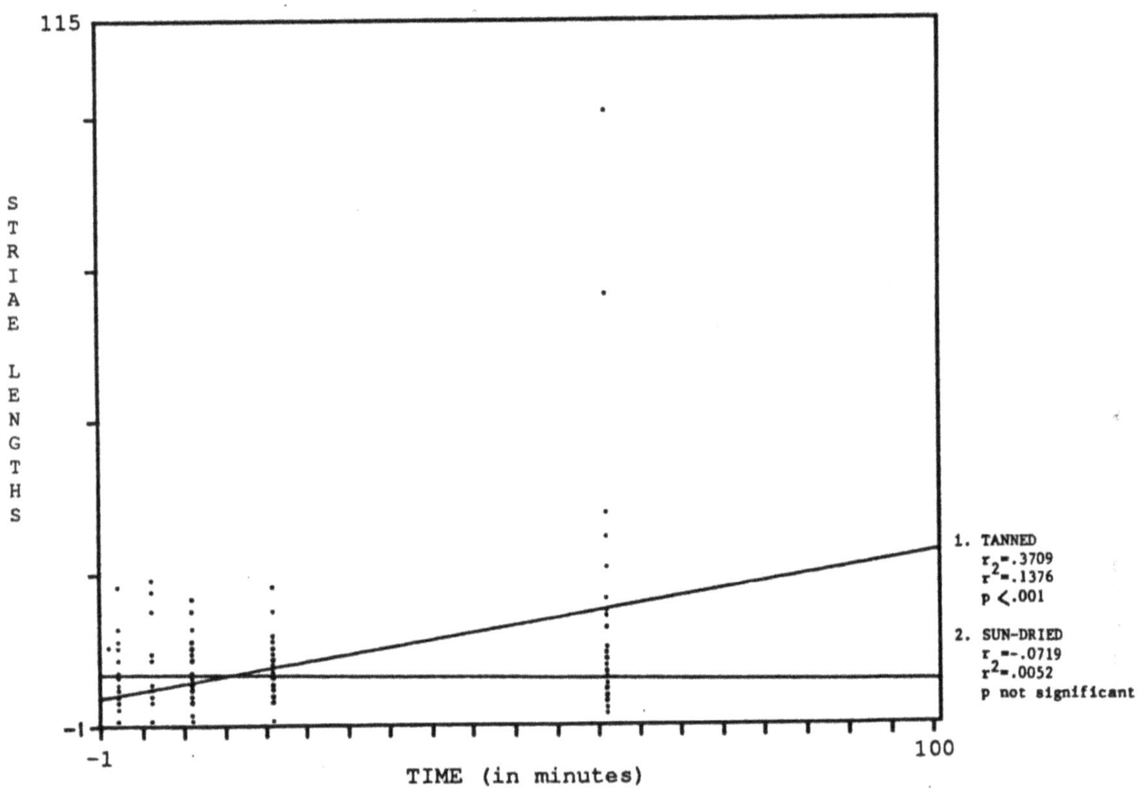

Fig. 6

Plate 25. Ventral Side of a Quartz Tool after Scraping Fresh Hide. Light Microscope 200X

Plate 26. Ventral Side of a Quartz Tool After Scraping Sun-Dried Hide. Light Microscope 200X

Plate 27. Quartz Edge After Scraping Tanned Hide Light Microscope 200X

Plate 28. Quartz Tool Used for Hide Scraping. Edge Upturned to show Attrition Light Microscope 200X

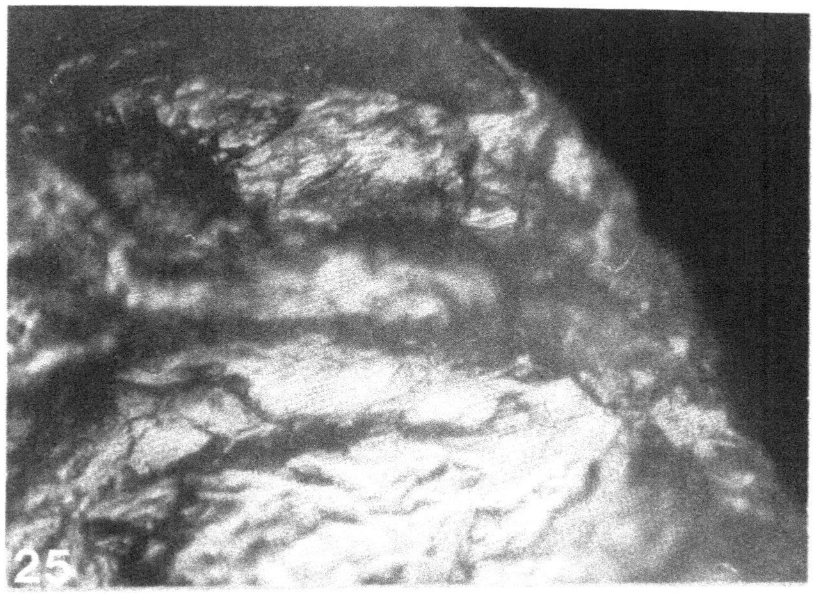

25

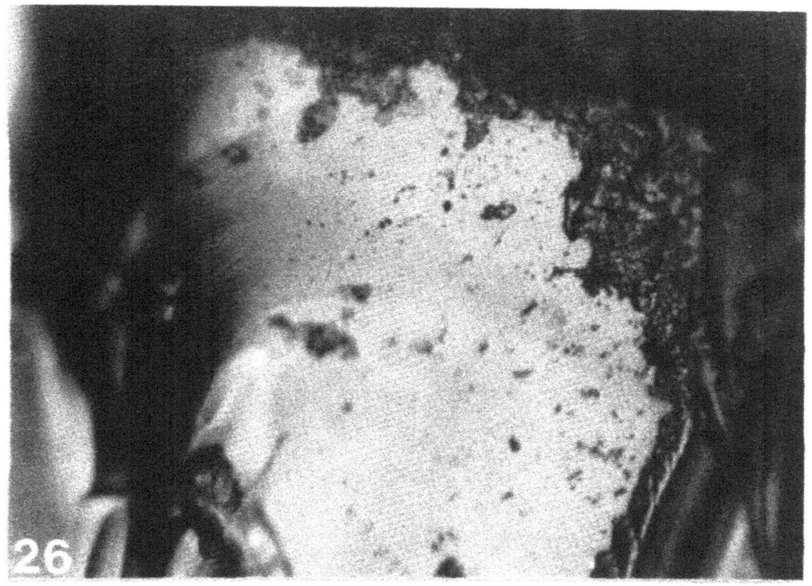

26

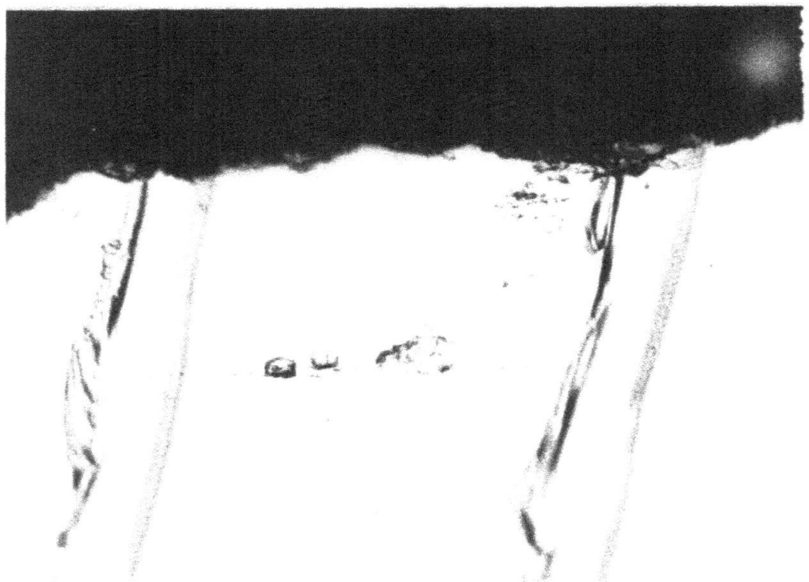

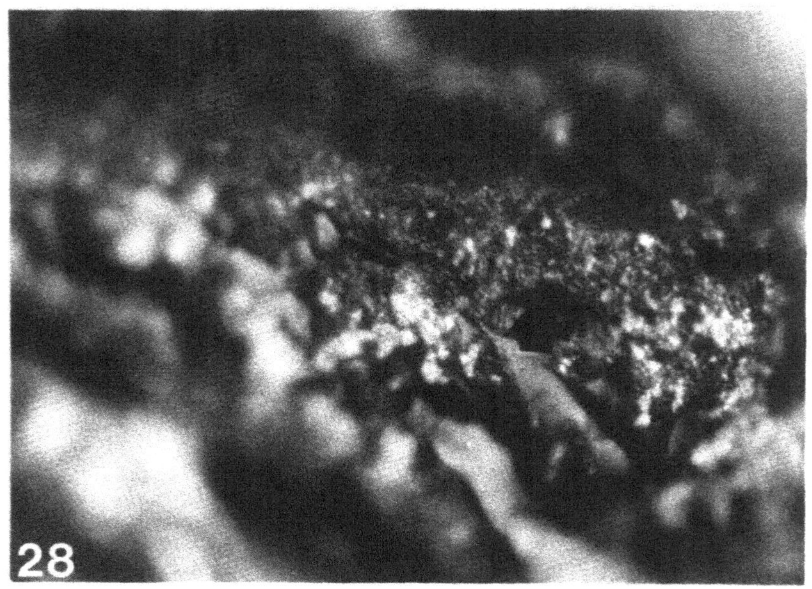

28

Plate 29. SEM Photo. Used Quartz Edge showing Edge Rounding from Attrition 500X

Plate 30. Flint Edge after Scraping Fresh Hide Light Microscope 200X

Plate 31. Flint Edge after Scraping Sun-Dried Hide

Plate 32. Quartz Edge After Scraping Re-Wet Hide Light Microscope 200X

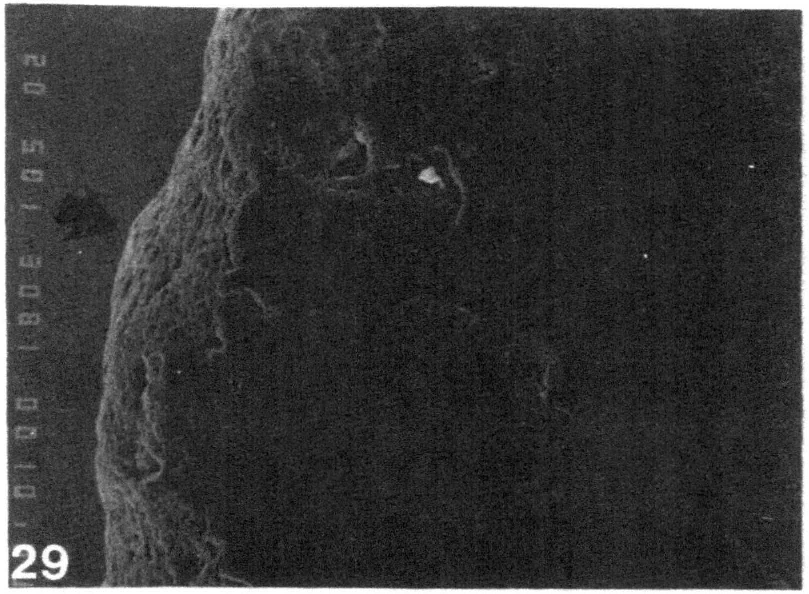
29

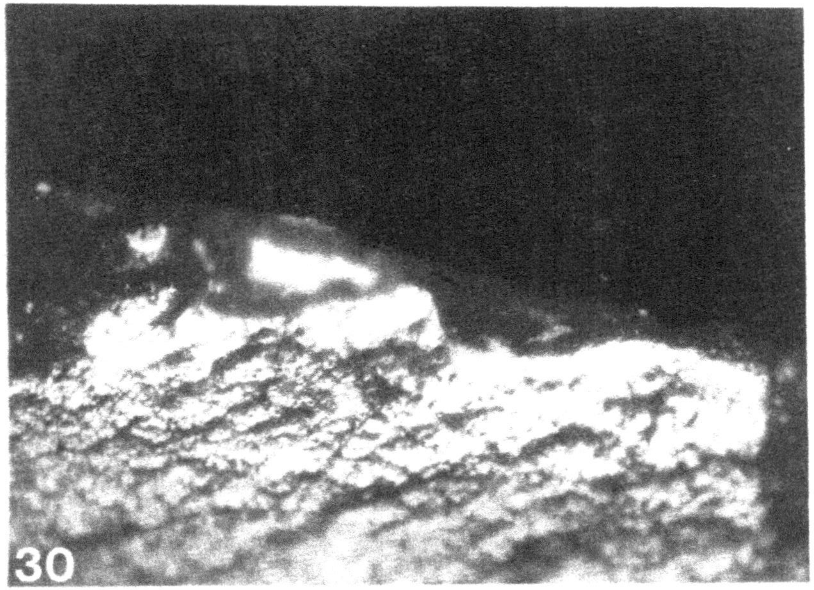
30

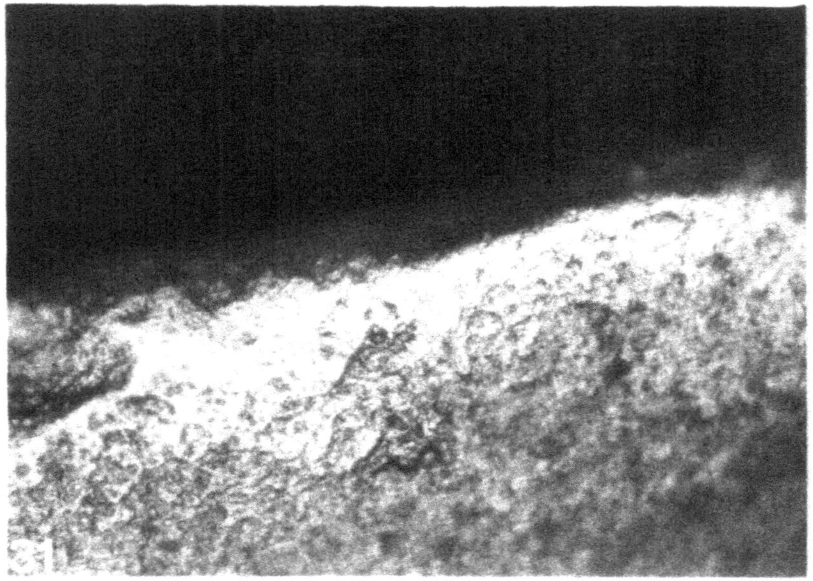

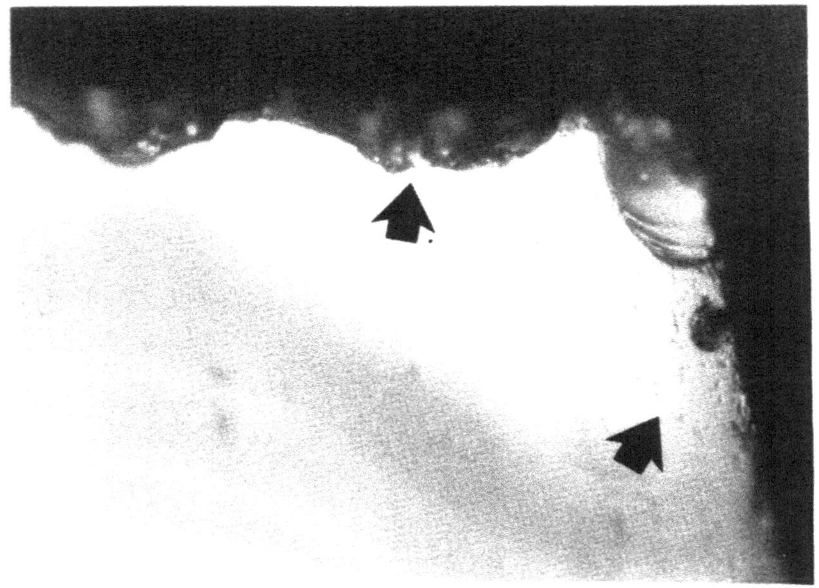

Plate 33. Quartz Edge after Scraping Fresh Hide
with Volcanic Ash Added
Light Microscope 200X

Plate 34. SEM Photo. Quartz Edge after Scraping Fresh Hide
with Volcanic Ash Added 200X

Plate 35. Crystal Quartz Edge used to Cut Beach Grass
Note Edge Damage. Light Microscope 200X

Plate 36. Crystal Quartz Edge After Cutting Beach Grass
Note Pits and Striae. Light Microscope 200X

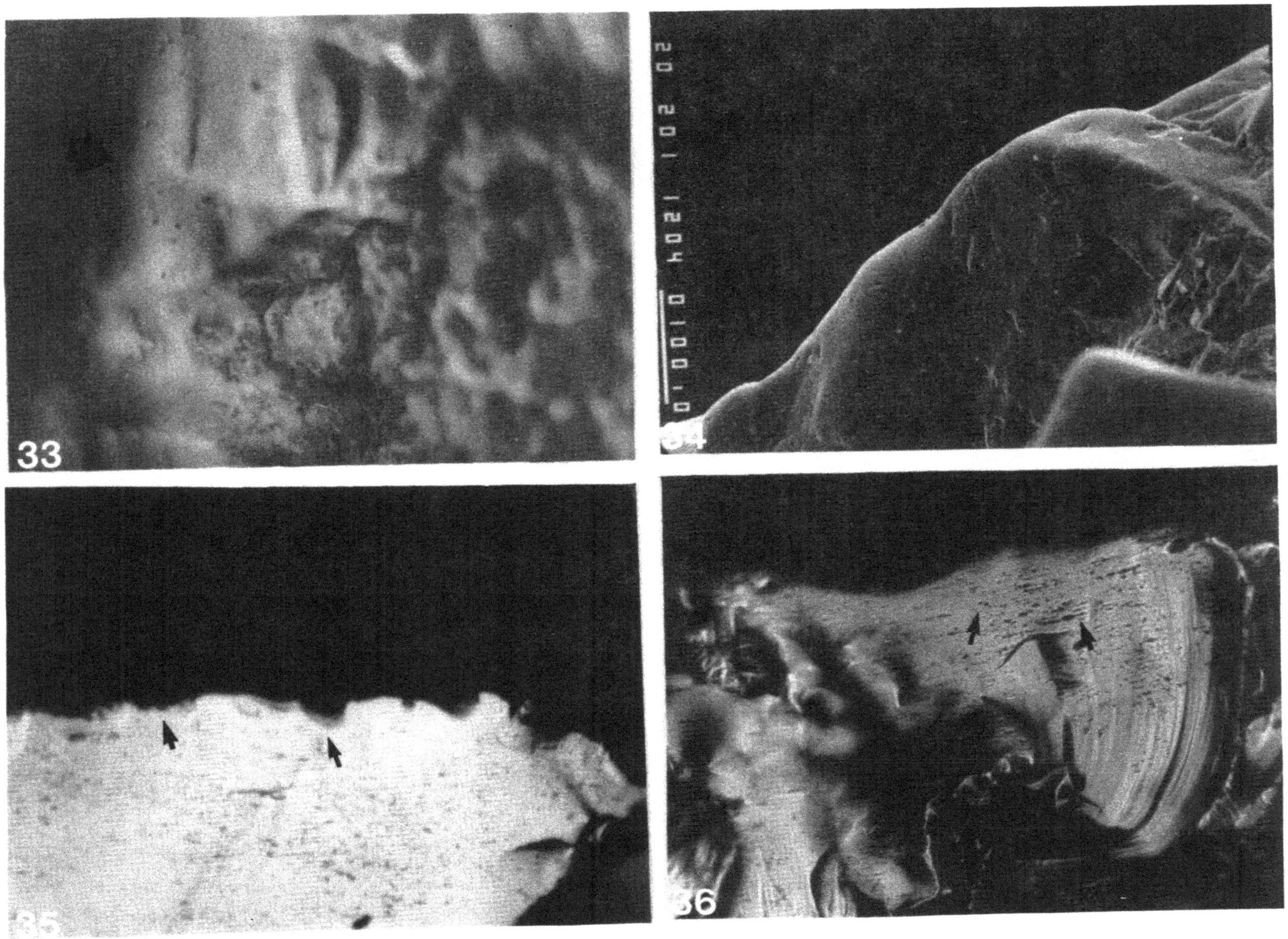
33
34
35
36

Chapter 4

Grass Cutting

Total Number of Flakes: 25

Total Number of Observations (5 per piece): 125

Number of Flakes used at Drake's Beach: 10

Number of Observations for Drake's Beach Flakes: 50

Number of Flakes used at Tilden Park: 11

Number of Observations for Tilden Park Flakes: 55

Types of Grasses Worked:

Drake's Bay Beach - *Ammophila arenaria* - beach grass

Tilden Park - *Agrostis alba* - bent grass

Avena fatua - wild oats

Avena sative - wild oats

Poa pratensis - Kentucky bluegrass

Scars - Edge damage in the form of scalar scarring was present intermittently along the quartz cutting edge as soon as use began. Half moon scars occurred only on those flakes used to cut beach grass for five minutes. For the sake of convenience, reference shall be made to Drake's Bay Beach grass as "beach" grass and grass from Tilden park shall be called "park" grass; the difference being that grass from Drake's Bay is long-stemmed and fibrous grass while the grass at Tilden Park is varieties of lawn grass run wild. As use time increased for all types of grass, the scalar scars became more numerous and overlapping. Edge damage was present on both sides of the cutting edges of all the tools. The stalks of *A. arenaria* are very brittle and hard compared with the various park grasses and subsequently created more edge damage along the cutting edges (Plates 6 & 35). This brittleness may play a part in the formation of pits as well.

Pits - These were visible after one minute of use on all types of grass and continued to be present throughout all time sequences. The flakes used for cutting beach grass, however, sustained more than twice the number of pits than those used on the park grasses (Table 5). This is shown more clearly when the experiments are examined individually (see Appendix 3). For example, after ten minutes of cutting the park grasses, a minimum of 2 and a maximum of 45 pits (mean=11 ± 13.5) were noted. After the same amount of time, a minimum of 2 and a maximum of 130 pits (mean=51.3 ± 45.1) were noted on the Drake's Bay pieces. Random flakes were examined microscopically prior to use and none showed any signs of pitting. Statistically, the number of pits found on both sets of tools correlates strongly with time (Fig

7). What is important, is that the pitting is microscopically distinct and easily seen. The "hard" data then substantiated what was first visually described (see Plates 36 & 37, and compare with the well developed silica polish on flint in Plate 38).

Numbers of Pits

	N	Mean	S.D.	Range
All Grass Working	105	56.55	84.63	0-500
Drake's Bay Grass	50	89.48	108.45	0-500
Tilden Park Grass	55	26.61	33.16	0-175

N refers to the number of attribute counts (5 per piece)

Table 5

Striae - Striae were not particularly numerous on any of the quartz tools used to cut grass, but there were significant statistical relationships formed between the numbers of striae found (Table 6) and the length of time a tool was used (refer to Fig.9). The numbers of striae on the 'beach' tools increased over time with only slightly greater statistical significance ($p<.01$) than the striae on the tools used for cutting 'park' grass ($p<.05$). The majority of all striae were narrow although after twenty minutes of use, one tool used for cutting beach grass (*A. arenaria*), sustained striae which were broad and shallow.

Striae lengths were measured at 200X with a micron scale using the light microscope. The mean range for lengths can be found in Table 7. The quartz flakes used on park grass show a longer mean striation length, while those used for cutting beach grass sustained almost twice the number of striae. Interestingly, striae lengths were statistically significant and correlated well with time but, in opposite directions. The lengths of striae on beach grass tools gradually increase over time, while those on park grass tools decrease over time (Fig. 10). The beach grass probably contained more abrasives (sand), than park grass so, one would expect not only more striae on those tools used for cutting, but also an increase in their length, as sand becomes caught in the abrasion tracks on the surface of the tool and is continually drawn back and forth. Striae on quartz used for cutting park grass became obscured or obliterated as polish formed more rapidly relative to the abrasive agents present. It is possible, although at present undocumented, that park grass contains more silica than beach grass which could explain a faster rate of polish formation. The phenomenon suggested by Mansur-Franchomme (1983) to account for the lack of striae on flint tools used to work hide may be operating here (refer to striae section in hide working chapter). The interraction between the plant silica and the surface of the quartz may create a "gel-like" state on the surface so, as a striation begins to form, it is immediately filled in by the surrounding silica which has not yet hardened.

The number of striae formed correspond to: the amount of moisture present in the soft plant material being worked, the amount of silica the plant contains and the quantity of abrasives present. Bettison (1985) has done experimental work cutting soft plant material with flint tools and concluded that there are actually **two** types of "sickle sheen" which form on the edges of the tools depending upon which stage in the life cycle of the plant it is at when harvested. She concludes that dry, or more mature plant material will produce an attritional polish, visible as smoothing and rounding of the edge, while green, moist plant material produces a rounding of the edge and an additional layer. Bettison describes this addition as "a distinct layer lifting off the surface of the tool" when viewed using the scanning electron microscope (Ibid:28). I was unable to document those results and found, a silica gloss more typical of that described for flint by Keeley (1980:61).

The quartz tools used to cut beach grass sustained more "artefacts" on their surfaces in the form of pits and linear grooves than did those quartz flakes used on park grass. This may be due to a a chemical alteration taking place on the surface of the tool during use and abrasion as suggested by Unger-Hamilton (1983, 1984).

Out of a total of eleven flakes used on park grass, 4 flakes had **no** striae on their ventral faces while one had no striae on its dorsal face but striae on the ventral face. Tools used on beach grass showed striae on both their dorsal and ventral faces. There was no correlation between the length or presence of striae and side of the flake (e.g. dorsal vs. ventral) which might give an indication of handedness.

Numbers of Striae

	N	Mean	S.D.	Range
All Grass Working	105	.83	1.66	0-10
Drake's Bay Grass	50	1.14	1.67	0-8
Tilden Park Grass	55	.56	1.61	0-10

N refers to number of attribute counts (5 per piece)

Table 6

Lengths of Striae

	N	Mean	S.D.	Range
All Grass Working	401	7.01	8.26	0-130
Drake's Bay Grass	262	6.99	5.56	0-30
Tilden Park Grass	139	7.05	11.76	0-130

N refers to number of striae

Table 7

Linear Grooves - Quartz flakes used for cutting beach grass had a slightly higher mean number of linear grooves than the combined average, while quartz tools used on park grass sustained many less (see Table 8, Plates 36 & 37). Both by direct observation and statistical plotting, it was clear that the number of linear grooves increased significantly over time (Fig. 8).

Numbers of Linear Grooves

	N	Mean	S.D.	Range
All Grass Working	105	10.19	12.57	0-75
Drake's Bay Grass	50	14.86	14.86	0-75
Tilden Park Grass	55	5.94	7.93	0-39

N refers to number of attribute counts (5 per piece).

Table 8

Ridge Attrition - Attrition of the ridges and high points did not increase as time of use increased, rather, its presence was sporadic, apparently dependent upon the microtopography of the flake's surface; a flake whose surface was irregular would be subject to attrition along the high points (ridges), whereas a flat surface (such as a crystal face), would not. On these particular sets of flakes there was only very minor ridge attrition.

Polish - A questionable lustre was seen after one minute of cutting park grass using quartz. After five minutes of use, the quartz flake used on park grass had a "metallic" looking polish along the worked edge and after ten minutes both the park and the beach tools showed the same "metallic" lustre. This continued upto and including one hour of cutting all grasses, with polish becoming increasingly luminescent and distinctive as cuuting time increased.

Edge Rounding - Rounding of the edge was first present on those tools used to cut park grass for five minutes and next, on beach tools after ten minutes of use. Unlike polish, edge rounding did not become more pronounced as time progressed, which may be due to simultaneous attrition of the edge as it is used and deposition of silica from the plant. Obviously, until definintive wotk on polish formation is accomplished, these hypotheses remain speculative.

Ridge Rounding - This was found on one of the quartz tools used to cut beach grass after ten minutes and was not noted again until after an hour's use, where it was clearly visible on both sets of experiments. Like ridge attrition, the amount of ridge rounding present is dependent upon the microtopographic features on the individual tool.

Surface Texture - After sixty minutes of use, the beach tools had a roughened surface texture caused by pitting in addition to silica polish, while the tools used for cutting park grass were much smoother.

Summary - In general, microwear traces seen on quartz flakes used to cut soft plant material are not so dissimilar from those described for flint (see Plates 37 & 38). Keeley, (1980:60-61) quoting Witthoft (1967) writes, "Witthoft listed the distinctive features of corn gloss as: (1) a very smooth, highly reflective surface, (2) a "fluid" appearance, (3) "filled-in" striations, and (4) "comet-shaped pits" ". On quartz, the only feature lacking is "comet-shaped pits". They do exist, but are encountered only after using quartz on bone or soaked antler.

The rough texture found on particular areas of the worked edges of the beach tools is due to pitting. The polish from all types of grass formed down along the functional edge and inside the damage scars as well. Although unable to document this photographically, I found a well-developed silica gloss to be "brighter" than the surrounding quartz surface, making identification relatively easy. Even so, identifying a plant polish on quartz may not be as straightforward as finding a well-developed plant polish on an archaeological piece of flint.

Regression: Pit Formation on Time for Grasses

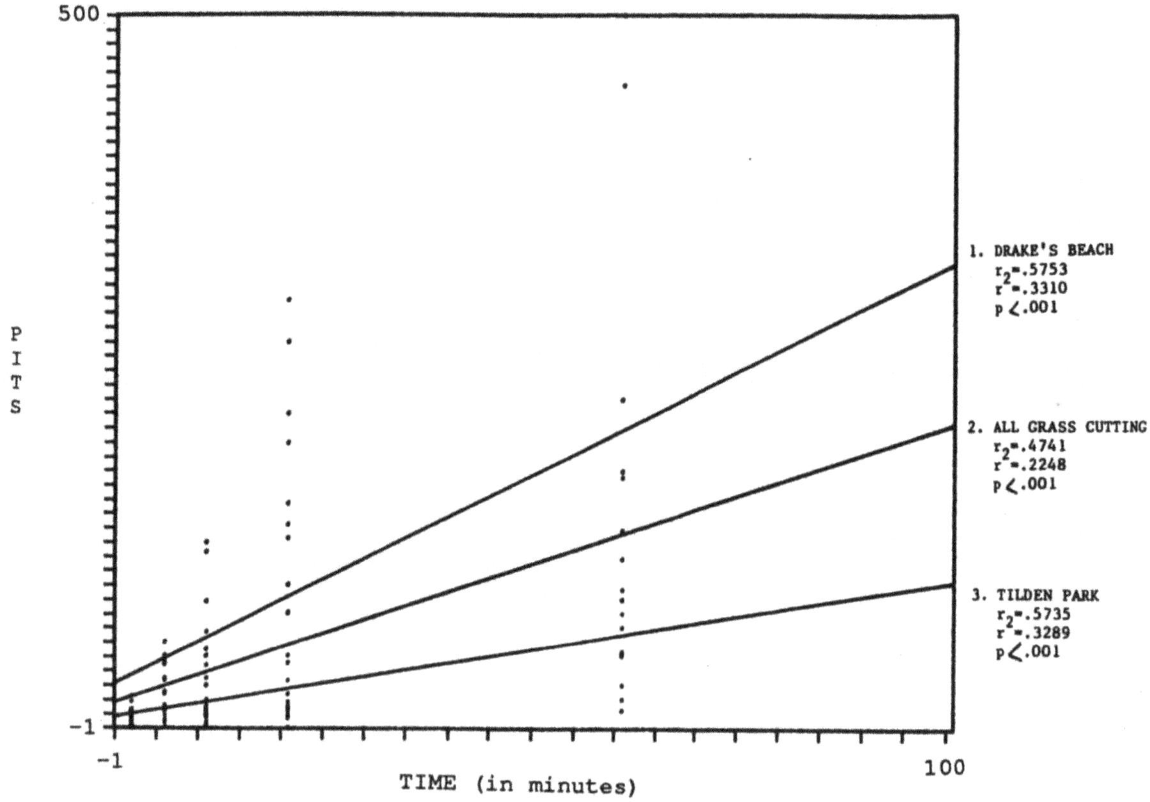

Fig. 7

Regression: Linear Groove Formation on Time for All Grasses

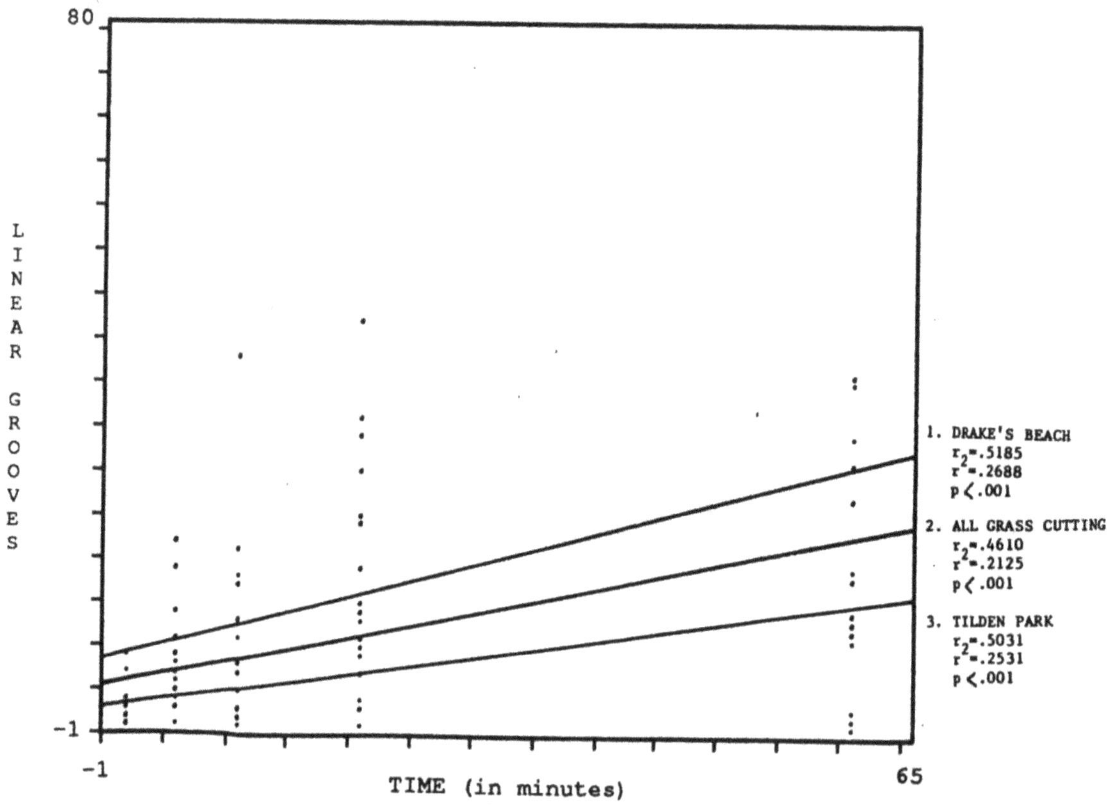

Fig. 8

Regression: Striae Formation on Time for All Grasses

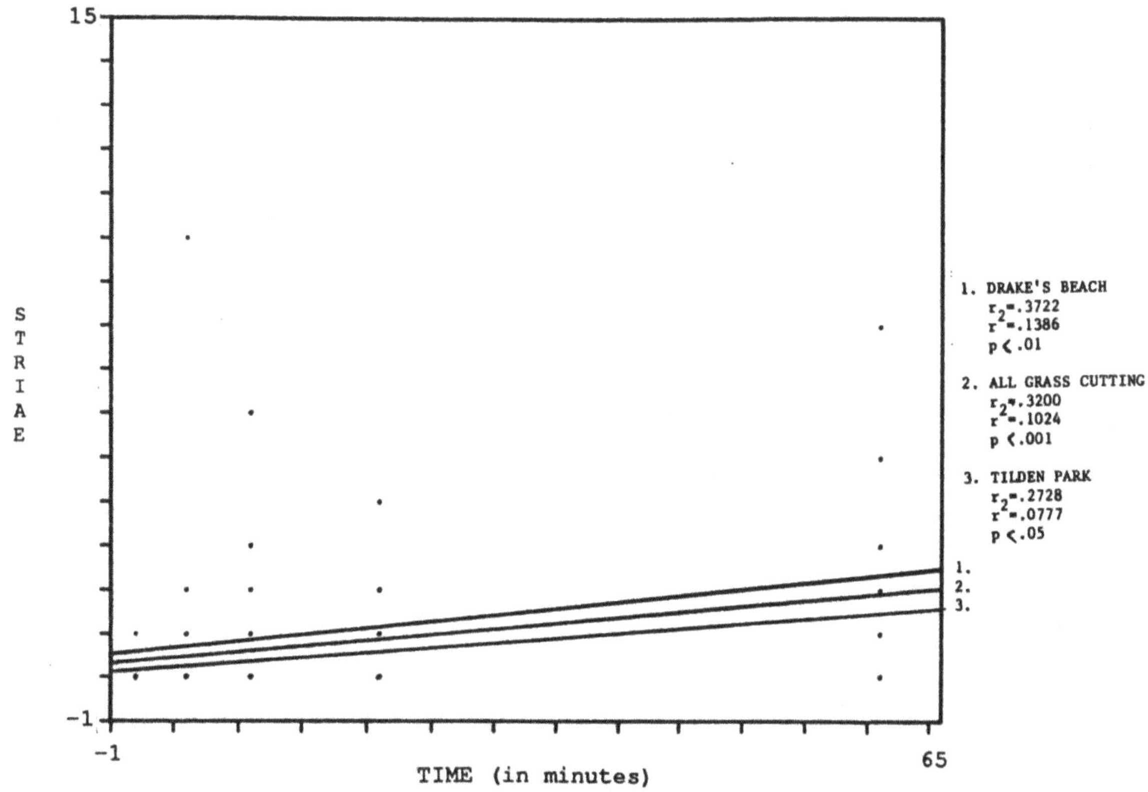

Fig. 9

Regression: Striae Length on Time for all Grass

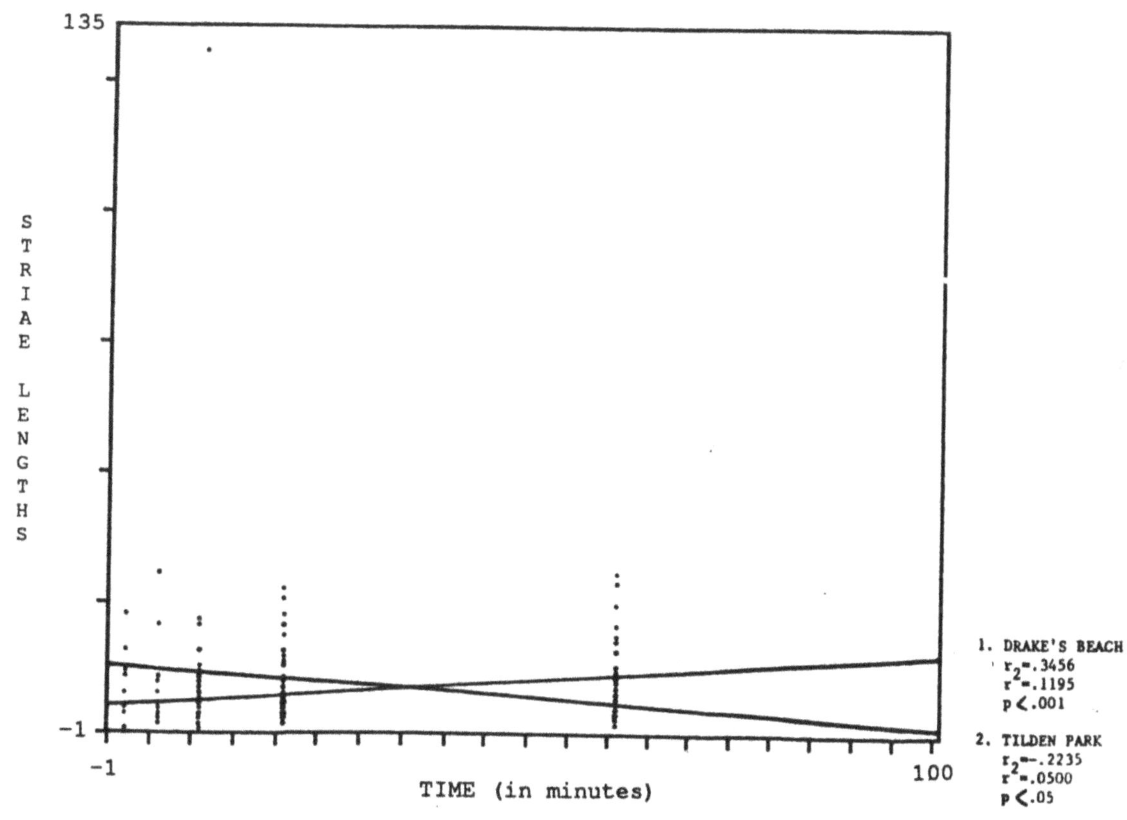

Fig. 10

Plate 37. Crystal Quartz Edge after Cutting Park Grass
Note Smoothness compared with Plate 36
Light Microscope 200X

Plate 38. Flint Edge After Cutting Park Grass
Light Microscope 200X

Plate 39. Wood Polish on Flint Light Microscope 200X

Plate 40. Surface of Quartz Tool after Sawing Oak
Note Smooth Surface and Attrition
Light Microscope 200X

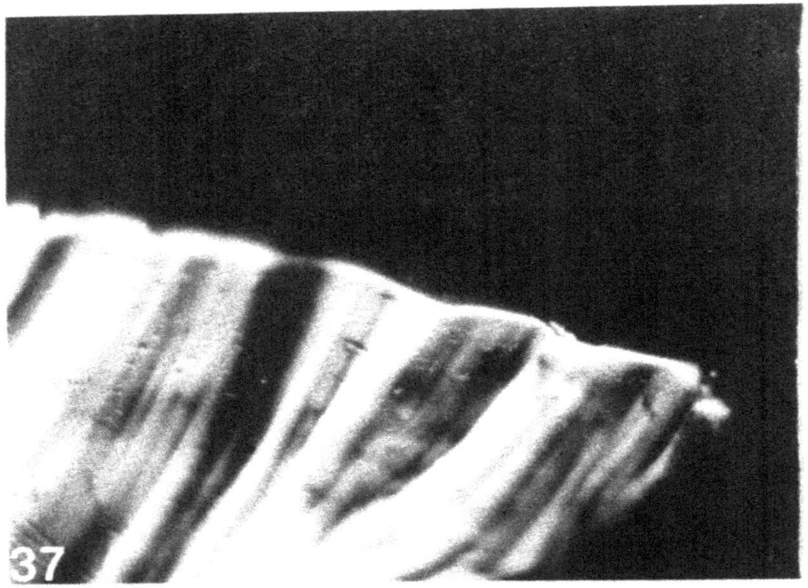

37

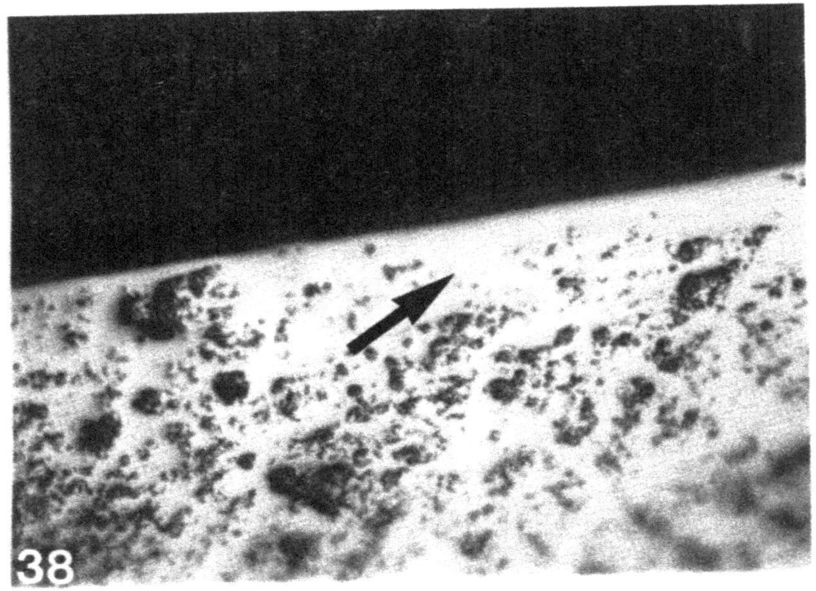

38

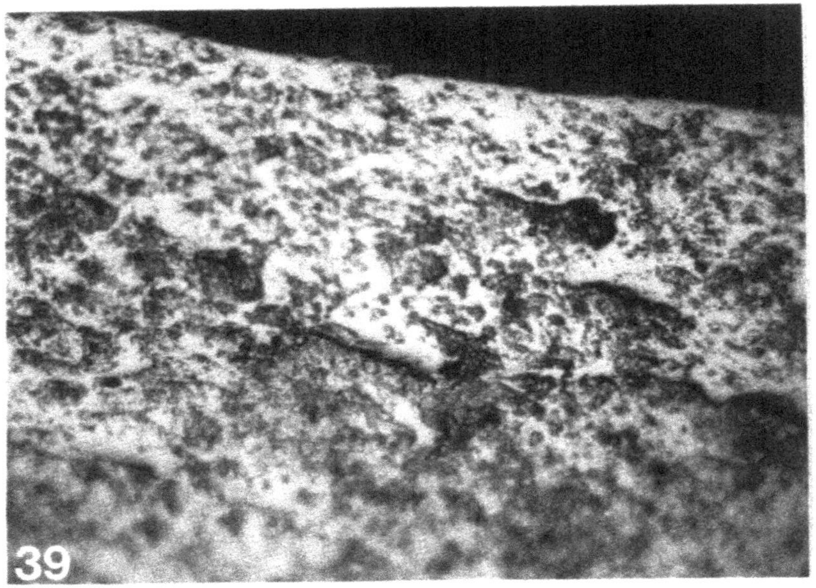

39

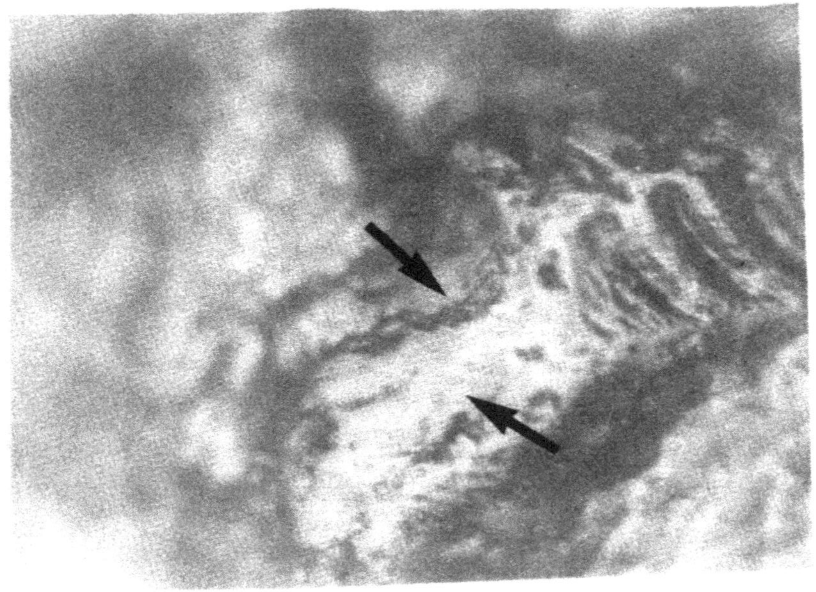

Chapter 5

Wood Working

Total Number of Flakes: 40

Total Number of Attribute Counts (5 per piece): 200

Number of Flakes used to Saw Wood: 25

Number of Attribute Counts for Sawing: 125

Number of Flakes used to Scrape Wood: 15

Number of Attribute Counts for Scraping: 75

Types of Wood Worked: Acacia melanoxylou - black acacia

Eucalyptus sp. (possibly lehmanii) - eucalyptus

Pinus sp. (possibly jeffreyi) - pine

Quercus sp. - oak

Taxus brevifolia - Pacific yew

Scars - Damage scars were found on both the dorsal and ventral sides of the quartz tools as a consequence of sawing and when scraping, scars formed, for the most part, on the side opposite that contacting the substrate.

Scalar scars were visible after one minute of use on all the woodworking tools, except the flake of vein quartz used for sawing fresh pine for one minute; it had no visible scarring along its the working edge. Scars may be present on the vein quartz tool but, undiscernable due to the reflective nature of the milky white, crystalline surface.

After ten minutes of sawing, step fractures became visible with occasional half moon breakages occurring after twenty minutes of use. These half moon breaks were most likely caused by the acute angle of the sawing edge on the quartz tool, and were not indicative of woodworking per se, as they were found only on one flake. The most prevalent and consistent scar type on all the tools was the scalar scar.

The scar patterns found on the quartz woodworking tools were similar to those described by Keeley for woodworking using flint tools (1980:35), e.g. edge damage lacked a consistent pattern or variation that could be definitively associated with woodworking.

Striae - The majority of striae produced by woodworking were broad and shallow. They became visible after sixty seconds of use and continued as time of use increased. Narrow, deep striae were present to a much lesser degree; seen only after scraping wood for twenty minutes. Keeley (Ibid:35) noted that striae were not common on flint tools used for wood-working, however, when present, they were broad and shallow. While I was unable to measure the widths of either types of striae, I differentiated between them when I could see the bottoms of the troughs at 200X using the light microscope. Measurement of the depths of the striae was not attempted.

The location and orientation of striae on stone tools are good indicators of the direction of use, along with linear grooves and, to a lesser degree, pitting. Sawing produced striae located close to the working edge and running parallel, or at low angles (less than 50 degrees), to it while scraping produced striae oriented more at right angles with the working edge.

In contrast to flint, quartz woodworking tools sustained many striations overall (Table 9), and their numbers bore a relation to the length of time a tool was used (Fig. 13). When statistical regressions were run on specific types of wood, only those flakes used to work oak or pine bore a significant correlation between time and the number of striae produced after sawing ($p<.001$ and $p<.05$ respectively, refer also to Fig. 17). The regression line for yew is potentially misleading as the sample size (N=10) is too small to be significant. The maximium number of striae on the quartz tools did not differ appreciably among the various kinds of wood, or the manner in which they were used (Table 9). Wood sawing and scraping activities gave a positive correlation between the number of striae and the amount of time a tool was used. The converse, that the number of striae indicates, to some degree, how long a piece was used unfortunately, does not hold true. People are using the tools, not machines. One person may use a tool with greater force, creating more minute edge damage and, perhaps, creating more striae; there may be more grit present during one activity compared with another, hence, more striae, and so on.

A regression analysis run on striae length and time showed both yew ($p<.001$), and eucalyptus ($p<.05$), to be significant for wood sawing, but to have no significance for wood scraping (Figs. 18 & 22). Both the length, and number of striations are undoubtedly influenced by the presence of dirt and grit in the bark and in the surrounding area where the activity is taking place (e.g. is the wood lying on the ground while it is being scraped?). The mean lengths for wood working striations on the experimental quartz tools are found in Table 10.

Numbers of Striations

	N	Mean	S.D.	Range
All Wood Working	200	1.20	2.09	0-15
All Wood Sawing	125	1.42	2.25	0-15
All Wood Scraping	75	0.84	1.72	0-8
Sawing Oak	25	1.12	1.50	0-6
Sawing Eucalyptus	45	1.37	2.51	0-15
Sawing Acacia	25	2.12	2.53	0-10
Sawing Pine	20	1.0	1.81	0-8
Sawing Yew	10	1.5	2.20	0-7
Scraping Eucalyptus	25	0.6	1.01	0-3
Scraping Acacia	25	0.28	1.37	0-7
Scraping Hardwood (unidentified)	25	1.64	2.24	0-8

N refers to numbers of attribute counts (5 per piece)

Table 9

Lengths of Striae

	N	Mean	S.D.	Range (in microns)
All Wood Working	1154	7.38	7.54	0-76
All Wood Sawing	946	7.53	7.92	0-76
All Wood Scraping	209	6.67	5.44	1-40
Sawing Oak	82	2.26	1.43	0-10
Sawing Eucalyptus	255	7.11	7.17	0-60
Sawing Acacia	437	9.29	8.58	1-70
Sawing Pine	32	3.68	1.84	0-10
Sawing Yew	140	6.79	8.28	0-76
Scraping Hardwood (unidentified)	181	5.69	4.07	1-32
Scaping Acacia	0	0	0	0
Scraping Yew	28	13.03	8.16	3-40

N refers to numbers of striations measured

Table 10

Pits - When all woodworking activities are combined and a regression analysis run on the data, the number of pits present correlates well with time (Fig. 11). The same is also true when sawing is considered, but no significant relationship emerges when scraping alone is examined. Sawing wood produced numerous pits on quartz tools, while scraping wood created relatively few pits (Table 11). The paucity of pits present after scraping wood was first noticed objectively while examining the tools microscopically, and was later substantiated quantitatively. Sawing yew left more pits on the surface of the tools than those used to saw other types of wood, and it was scraping hardwood (referring to broad-leafed as opposed to needled trees) that produced more pits on quartz then scraping other types of wood. It should be noted that only two quartz flakes were used to saw yew, and if the sample size was larger, the average number of pits might actually decrease.

Numbers of Pits

	N	Mean	S.D.	Range
All Wood Working	200	148.44	217.74	0-999
All Wood Sawing	125	186.12	254.31	0-999
All Wood Scraping	75	85.64	111.02	0-500
Sawing Eucalyptus	45	221.86	294.74	0-999
Sawing Acacia	25	254.72	279.26	35-999
Sawing Oak	25	102.84	154.22	0-350
Sawing Yew	10	280.9	319.01	0-999
Sawing Pine	20	76.7	114.16	0-350
Scraping Eucalytpus	25	48.84	116.28	30-400
Scraping Acacia	25	28.4	32.75	0-120
Scraping Hardwood	25	179.68	143.45	25-500

N refers to number of attribute counts (5 per piece)

Table 11

Linear Grooves - I hypothesized that the number of linear grooves on the surface of quartz tools used for working wood should decrease as use time increased because, the edge of the tool would stabilize and microchips would no longer break off the edge and be dragged across the surface. This, however, did not prove to be the case (for counts and times see Appendix 3). While the edge appeared to stabilize, it seems there was always some microscopic breakage of the edge during use.

Linear grooves were not common features on woodworking tools but, were more numerous on those tools used for wood sawing (Table 12). They appeared after one minute of use and were present after that on all tools regardless of the action or length of use.

Numbers of Linear Grooves

	N	Mean	S.D.	Range
All Wood Working	200	26.79	33.73	0-250
All Wood Sawing	125	31.42	39.45	0-250
All Wood Scraping	75	19.06	19.63	0-45
Sawing Oak	25	20.36	25.54	0-100
Sawing Eucalyptus	45	37.4	46.64	0-250
Sawing Acacia	25	50.2	42.24	8-200
Sawing Pine	20	9.2	11.15	0-40
Sawing Yew	10	26.09	26.09	0-95
Scraping Eucalyptus	25	21.92	21.06	0-72
Scraping Acacia	25	9.32	13.57	0-60
Scraping Hardwood	25	25.96	19.44	5-80

N refers to numbers of attribute counts (5 per piece)

Table 12

Sawing oak or pine showed a significant statistical correlation between the number of linear grooves present and the amount of time a tool was used ($p<.05$; $p<.001$ respectively), and scraping eucalyptus and acacia were statistically significant for linear grooves and time as well. As seen with striations, the regression line for yew looks "good" on the plot (Fig. 16), but should not be considered significant due to the small sample size (N=10).

The number of linear grooves found on each tool result from more than just physical contact between the quartz tool and the wood rather, their formation is more likely due to a combination of various factors including: stability of the tool edge (does it break into small pieces which are dragged across the tool's surface?); the action (does scraping produce less linear grooves than sawing?); and the relative hardness and density of the wood used. Controlled testing of this hypothesis with a larger sample size than used here, could produce some extremely interesting results.

Polish - The attributes of wood polish on flint as described by Keeley (1980) are distinctive; they have a consistent appearance regardless of the type or hardness of the wood, whether it is seasoned or fresh, or the manner in which it was worked. Keeley noted for flint, and it appears to be the same for quartz, that dense and seasoned woods create less polish than unseasoned or less dense woods.

Wood polish on flint is bright, smooth in texture and once visible, is found,

> "commonly gently curved or domed on the high points of the microtopography... After prolonged use, it will show gentle undulations on its surface with the lines of the "troughs" and "crests" running in the general direction of use" (Ibid:35).

When the surface of the flint is rough, the polish assumes a recticular pattern (Plate 39) as it forms on the high points first.

I found visible wood polish on the experimental flint tools occasionally after one minute of use, and certainly after ten minutes. Of course, identification of a polish usually becomes easier the longer a tool has been used as more polish forms over time. The attributes present were the standard features described above.

On the experimental quartz tools, however, wood working produced a different polish. Polish was visible after sawing pine for one minute but, when used for longer periods of time on any other type of wood, its presence was inconsistent and unpredictable upto and including one hour of scraping or sawing. The presence of polish after sawing pine for such a brief time may be caused by the amount of silica and moisture present in the wood. This has not been investigated thoroughly as yet.

When present on quartz, polish appeared bright and metallic along the working edge of the tool at 100X under the light microscope. After twenty minutes of use the polish took on a fluid appearance. It looked equally bright for both sawing and scraping, however, specific measurement of this brightness was not attempted due, in part, to the high degree of reflectivity of the quartz surface itself. It was more easily recognizable when associated with striae, pits and linear grooves. The quartz piece whose polish most closely resembled a "typical" wood polish on flint (bright, smooth and undulating), was a flake of vein quartz used to saw oak for an hour (Plates 40 & 44). It was interesting (and unfortunate), that there was an overlap of polish attributes between sawing wood and sawing antler at a certain stage in polish development. This caused a wood polish, such as that seen after sawing yew for 10 minutes, to resemble closely wear produced after sawing soaked antler for the same amount of time (Plates 42 & 43). The best prospect for an easily identifiable wood polish on quartz would be a tool used for a <u>long</u> period of time to saw or scrape a soft, freshly cut piece of wood containing lots of silica. Many microwear analysts agree that the character of a wood polish is affected by its moisture and silica content (see Moss 1983b:152).

<u>Surface Texture</u> - After five minutes of use, surface roughness was sporadic, lacking any apparent patterning or predictability. Roughness occurred in conjunction with smoothing, therefore, the presence of one does not preclude the presence of other, and both can be assumed to result from abrasion. High and low-lying areas of the microtopography were equally effected (Plate 44).

<u>Ridge Attrition and Ridge Rounding</u> - Attrition of the ridges

differs from the surface texture described above because it was found only on the high points and was consistently rougher (nubbly in texture) than that found on the quartz surface. After one hour of sawing or scraping wood, all the quartz tools showed varying degrees of ridge attrition associated with occasional ridge rounding. Ridge attrition occurred earliest on those flakes used for scraping. It is a remote possibilty that those flakes used for scraping had more micro-protrusions on their surface or that scraping produces more wear on the tool. Of course, a combination of both factors is a possibility as well.

Smoothing and rounding of the ridges was not seen earlier than one hour of scraping but was noted after twenty minutes of sawing.

Edge Attrition, Rounding, and Smoothing - Edge rounding should not be equated with smoothing; rounding produced a gentle and regular curving of the edge by blunting but, may leave a rough surface texture. Smoothing, on the other hand, results in an even surface where there is no roughness or projections visible.

Attrition occurred quickly and was present, to varying degrees, on all the experimental woodworking pieces of quartz. Severe attrition resulted in marked edge rounding although this did not become appreciable until more than twenty minutes of sawing or after one hour of scraping wood. Sawing wood using quartz resulted in more smoothing of the edge of the tool than did scraping, even though both attrition and smoothing were present (Plates 41 & 45).

Many of the flakes used in this "set" of wood working experiments were made from vein quartz. Due to the irregularity of the surface, crypto-crystalline nature of the raw material and the corresponding variability in surface textures, it was difficult to accurately identify and assess, in all cases, the edge and ridge rounding as well as the changes in surface textures which were due to use.

Summary - The appearance of wood polish on quartz was to variable in its occurrence but, when present, was bright, having a metallic luster and was associated with pitting and linear grooves. Difficulty in the identification of a wood polish arose when there was no polish visible, which left only pits, linear grooves and occasional striae as markers of use. The patterning of edge damage was not unique to working wood and overlap of wear attributes other than polish exists at certain stages of development. Therefore, a positive identification of a wood polish is more likely if polish is present along with pits, linear grooves and striations.

Regression: Pits on Time for All Wood Working

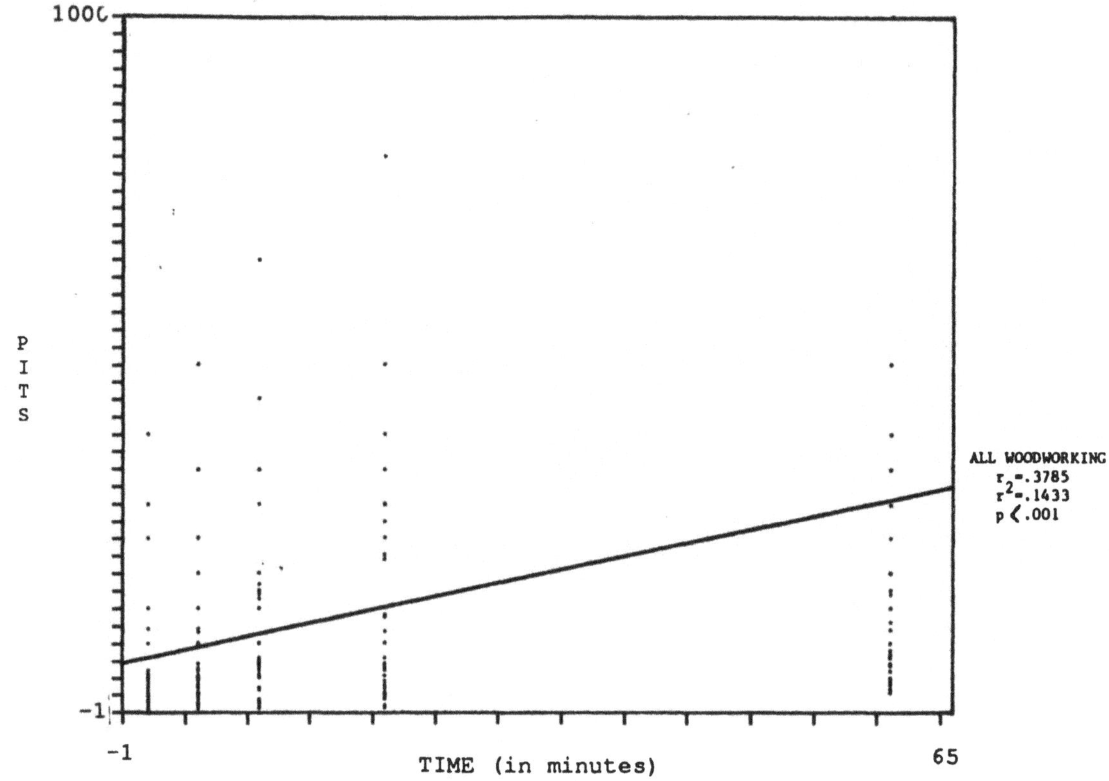

Fig. 11

Regression: Linear Grooves on Time for All Wood Working

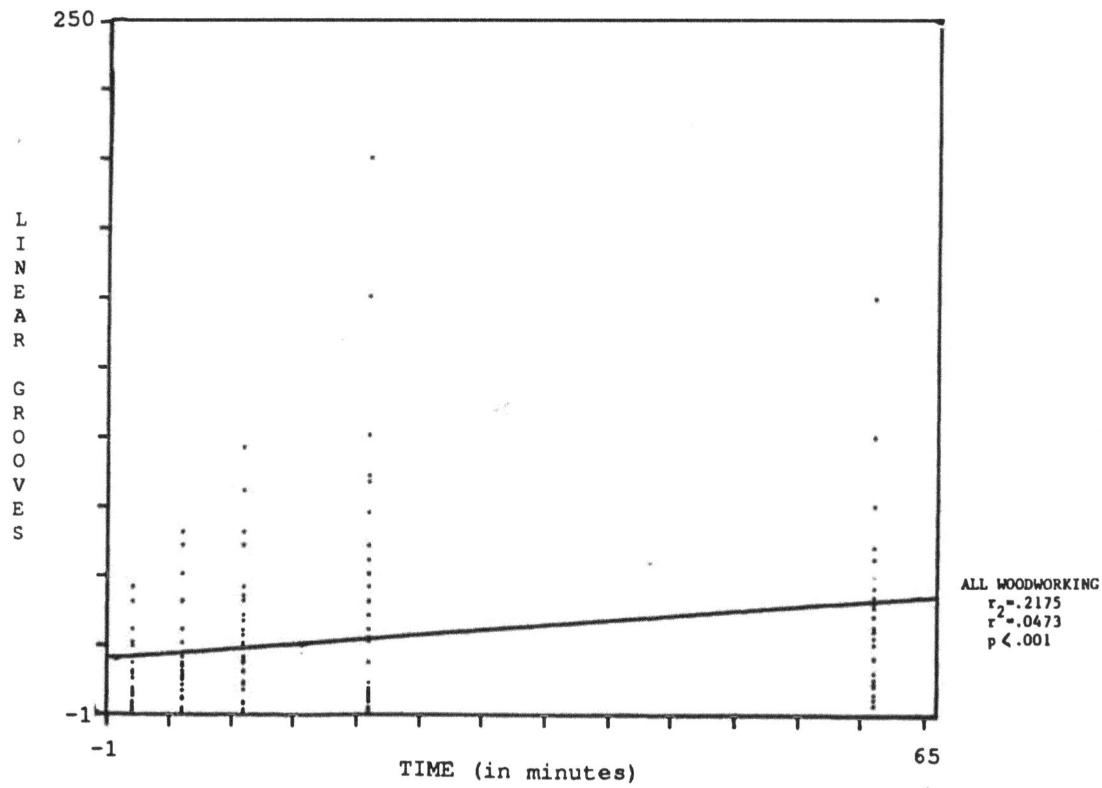

Fig. 12

Regression: Striae on Time for All Wood Working

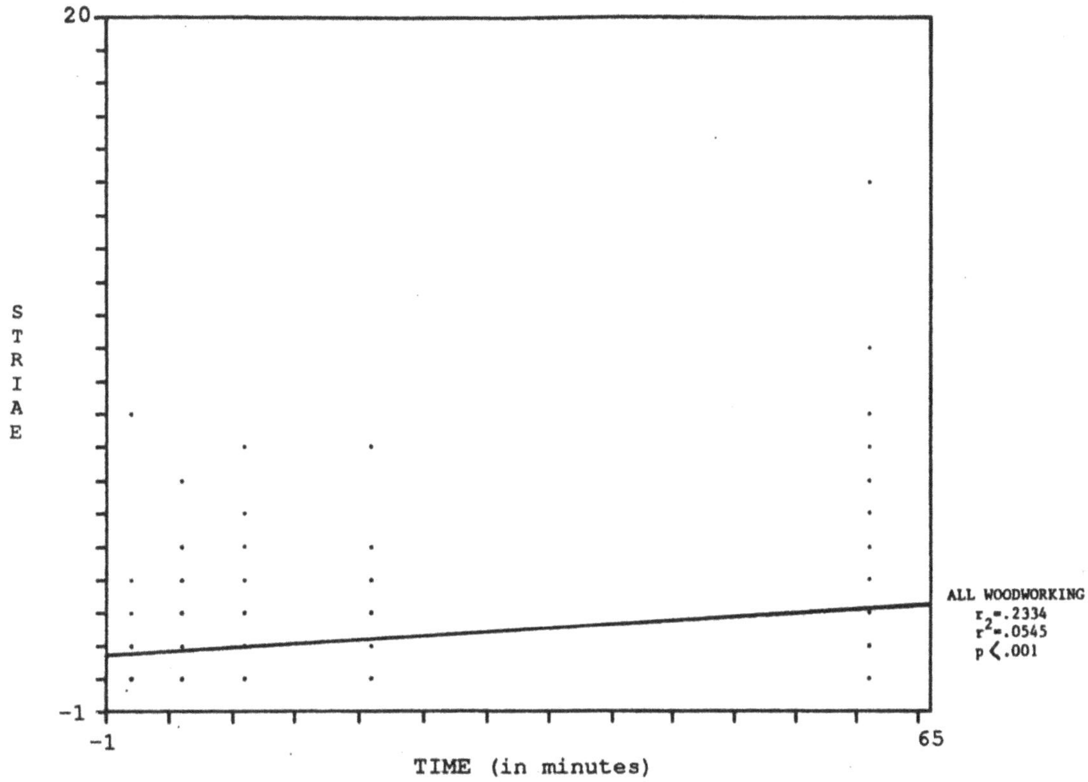

Fig. 13

Regression: Striae Lengths on All Wood Working

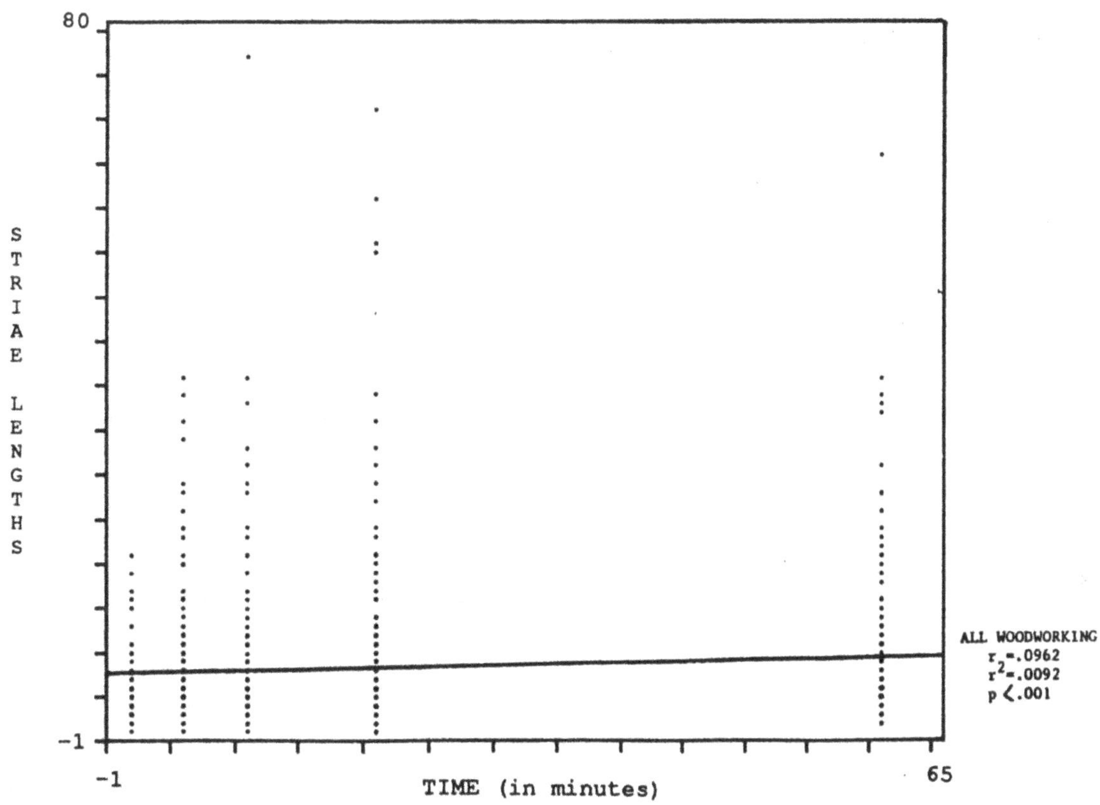

Fig. 14

Regression: Pits on Time for Wood Sawing

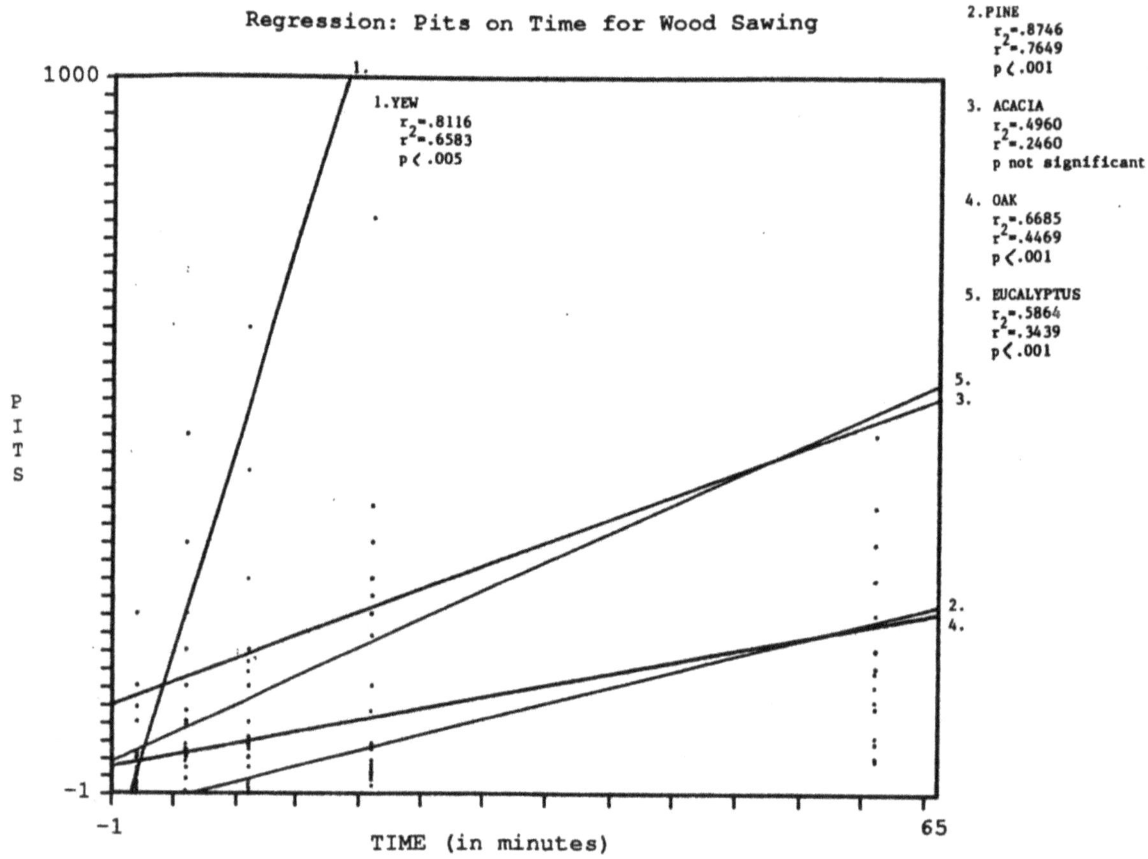

Fig. 15

Regression: Linear Grooves on Time for Wood Sawing

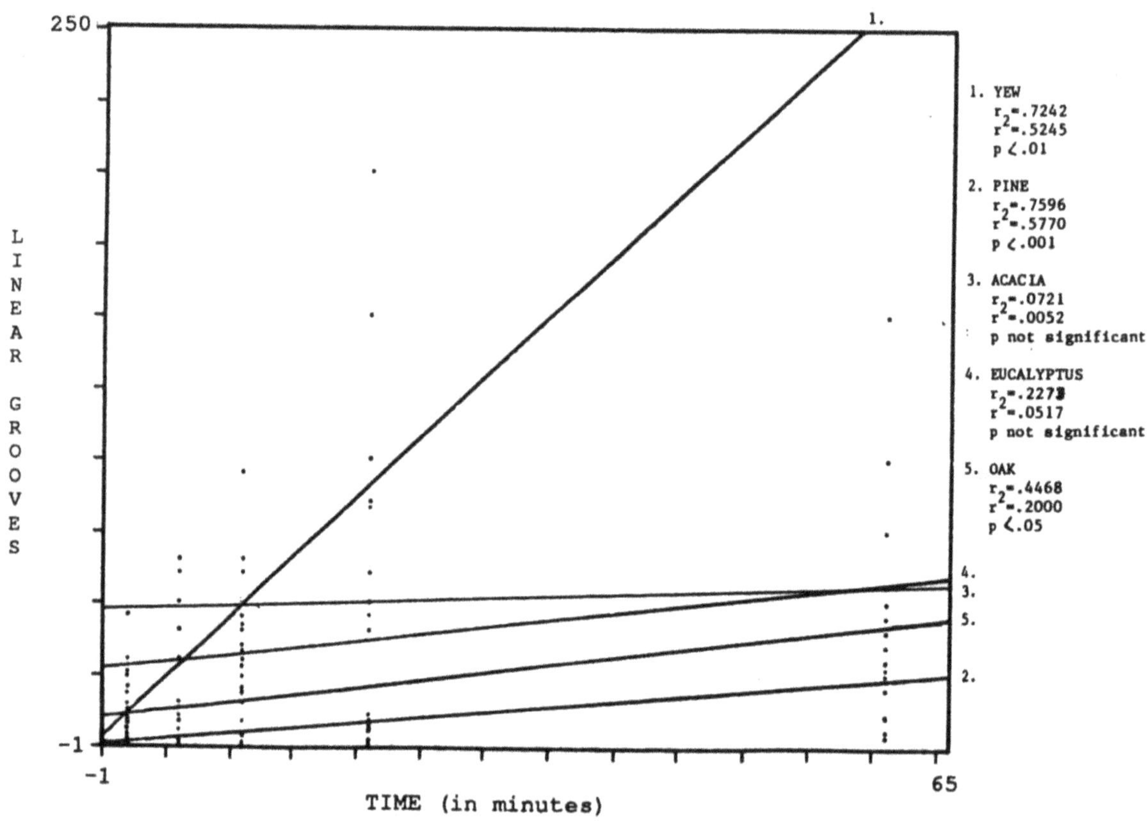

Fig. 16

Regression: Striae on Time for All Wood Sawing

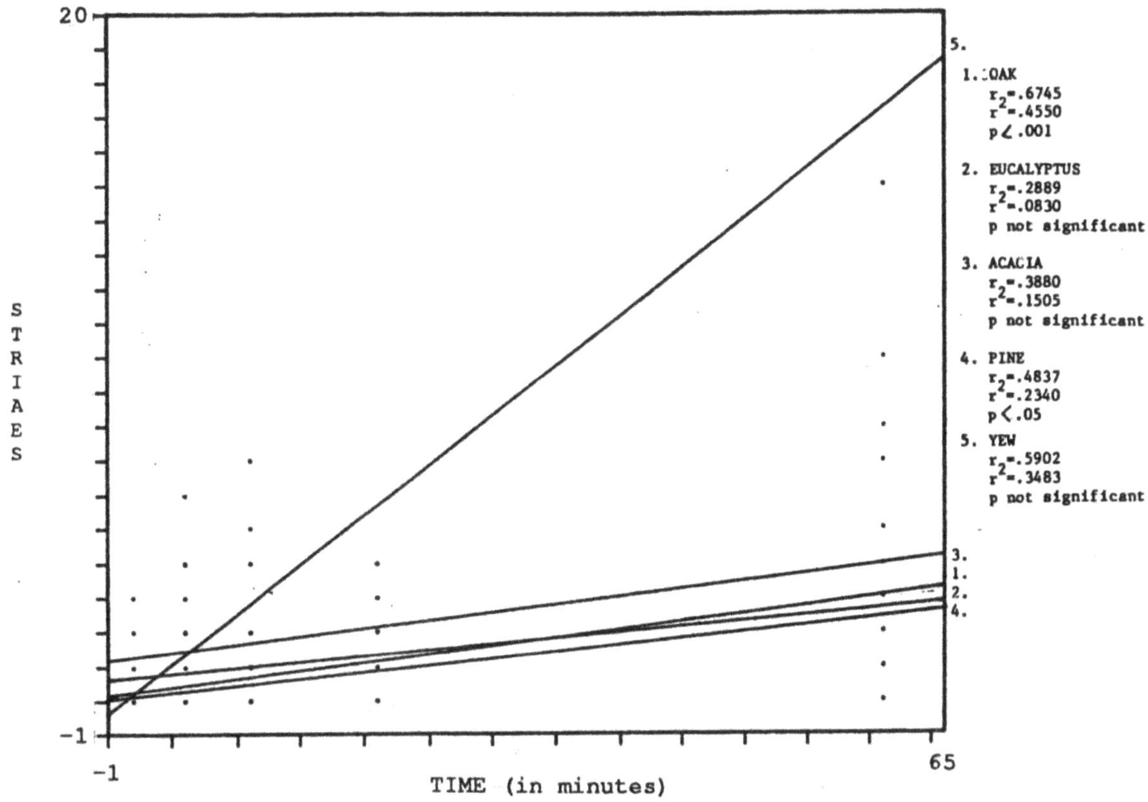

Fig. 17

Regression: Striae on Time for All Wood Sawing

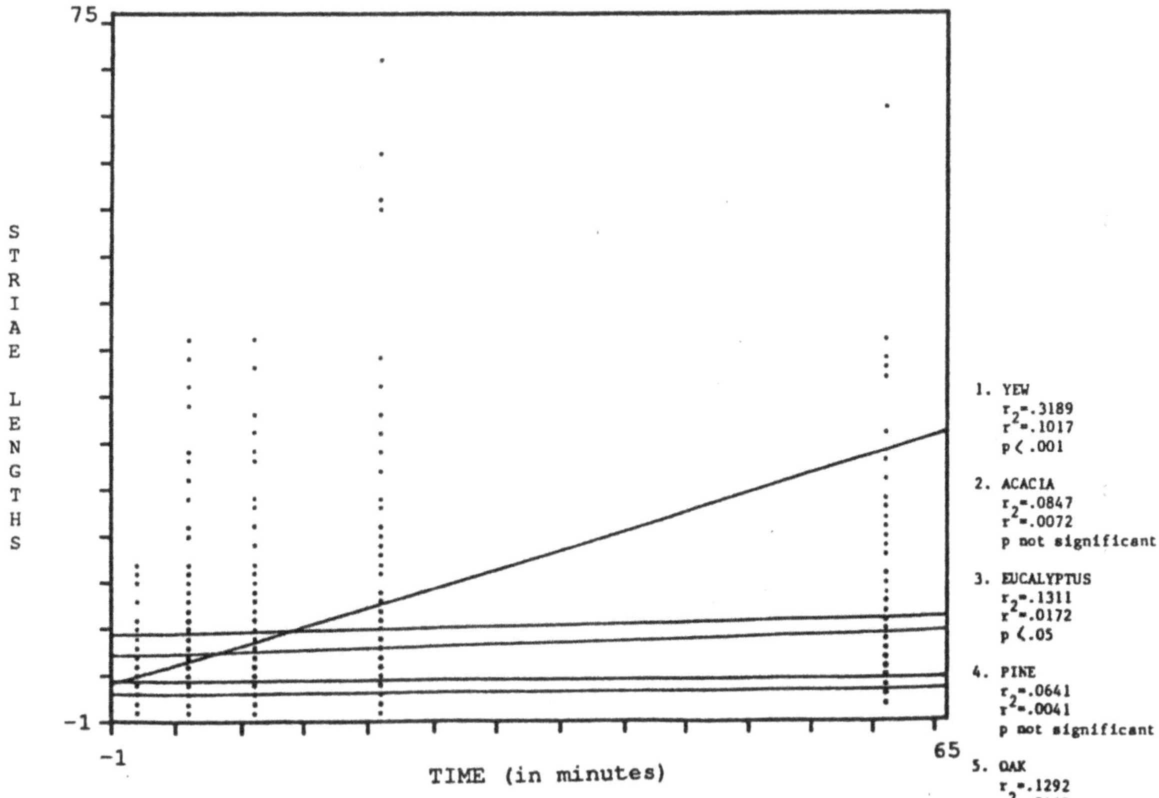

Fig. 18

Regression: Pits on Time for All Wood Scraping

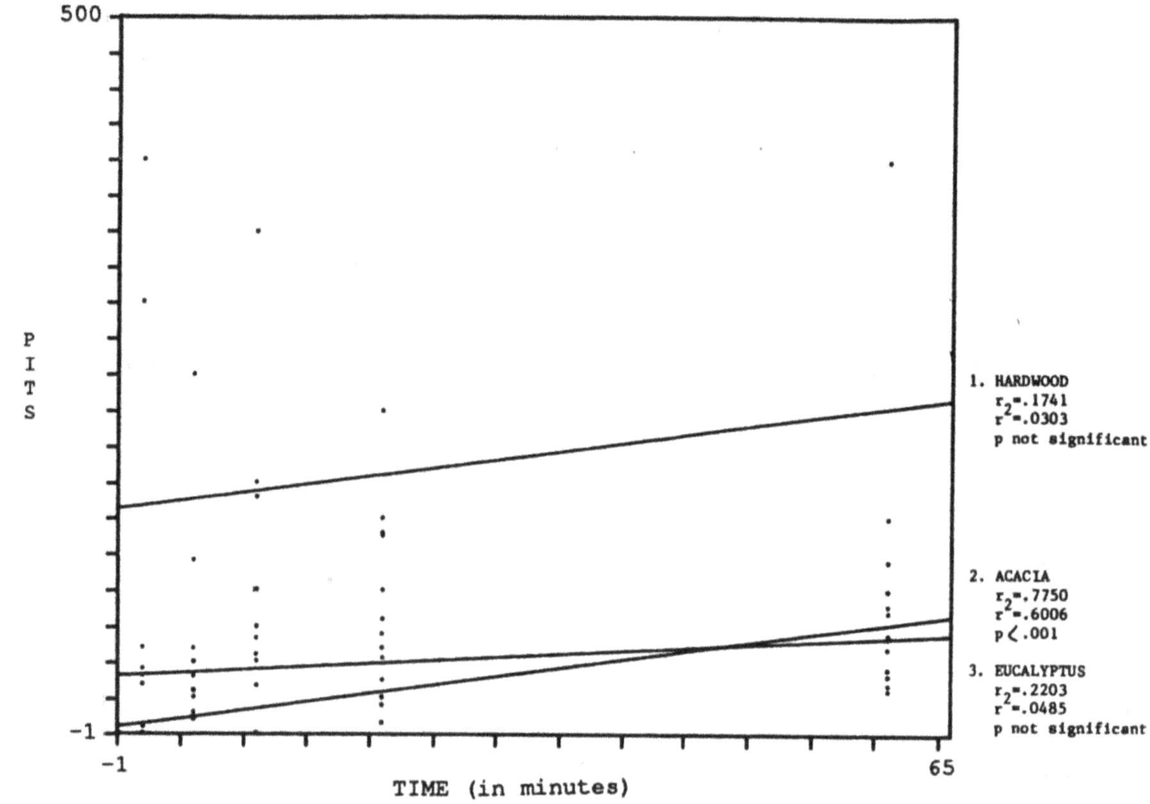

Fig. 19

Regression: Linear Grooves on Time for All Wood Scraping

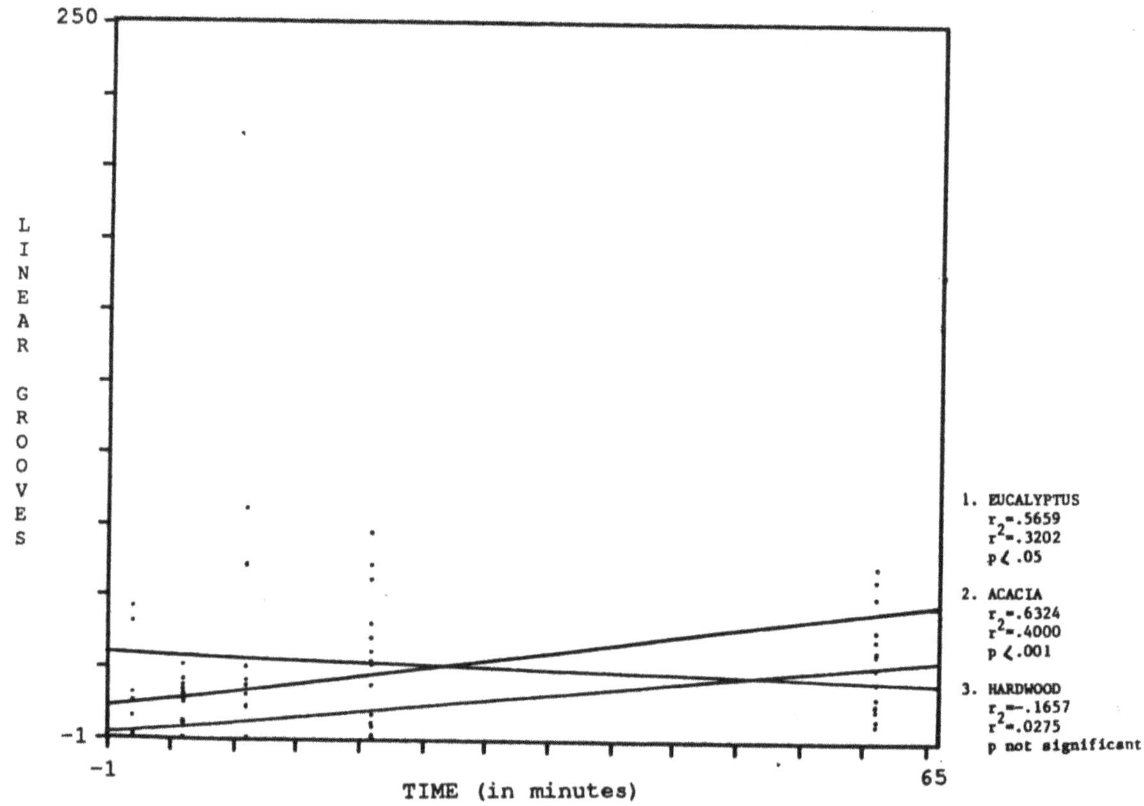

Fig. 20

Regression: Striae on Time for All Wood Scraping

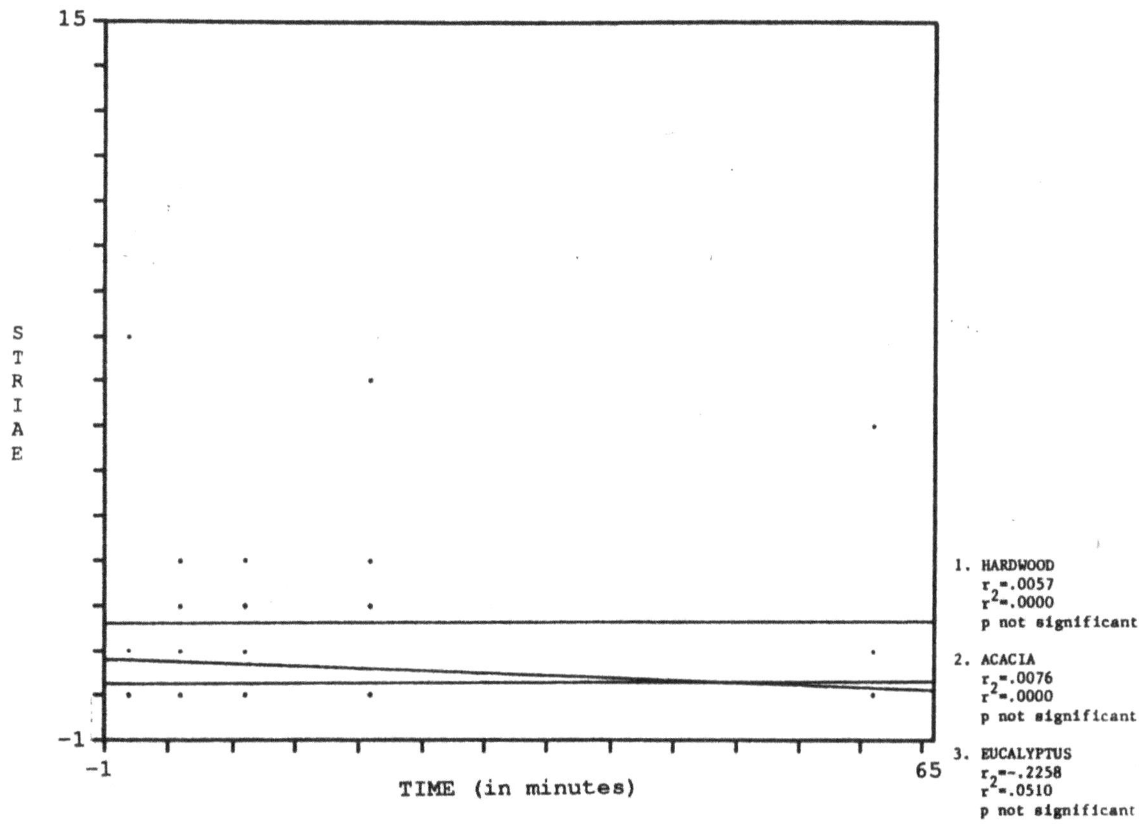

Fig. 21

Regression: Striae Lengths on Time for All Wood Scraping

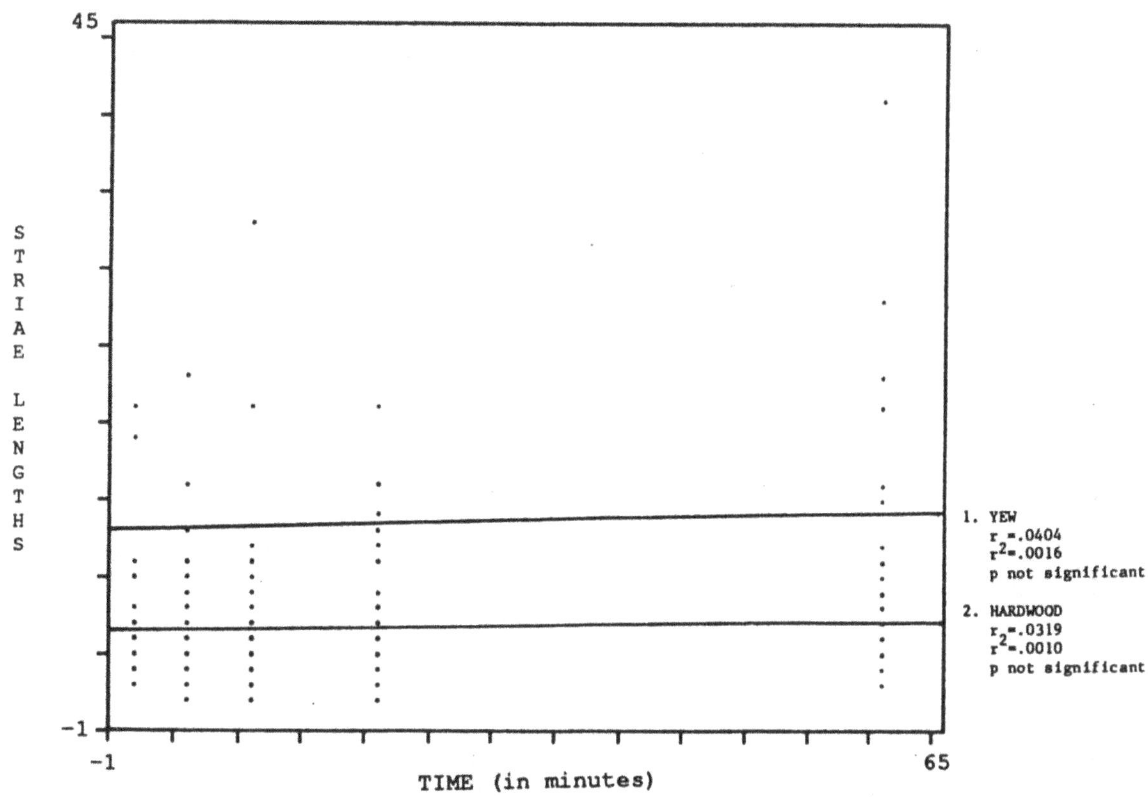

Fig. 22

Regression: Striae on Time for All Wood Sawing and Scraping

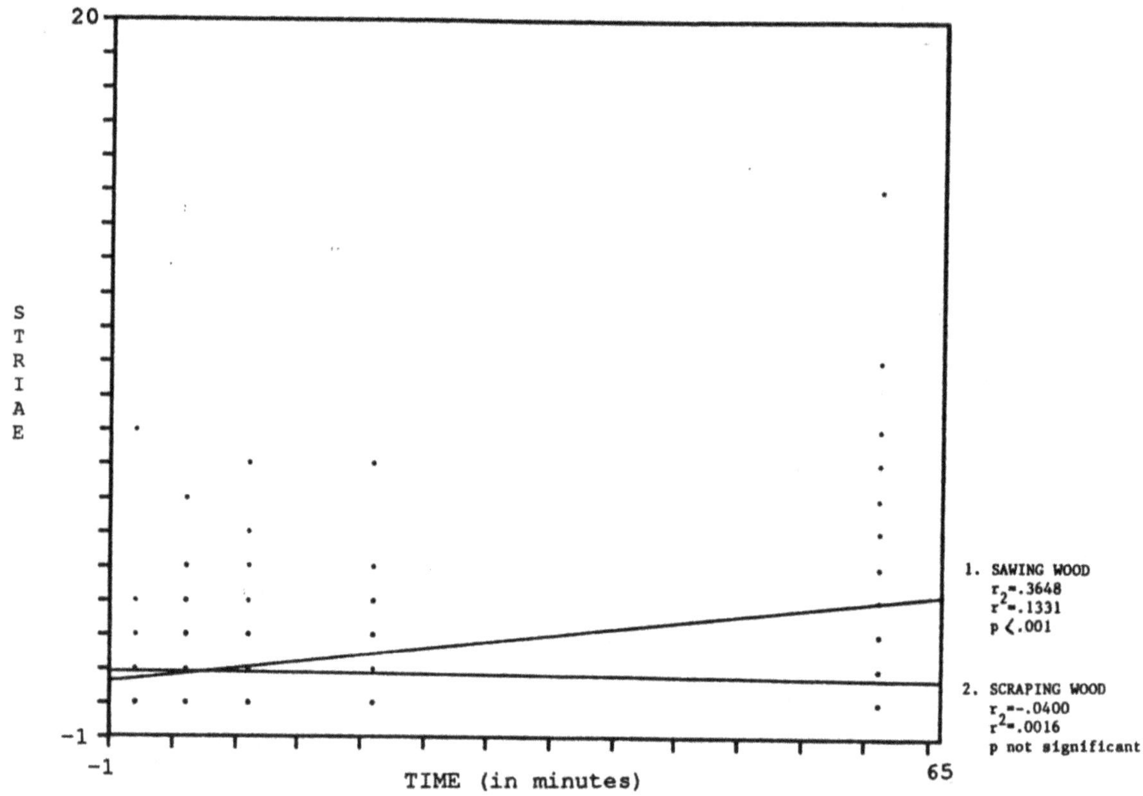

Fig. 23

Regression: Striae Lengths for Wood Sawing and Scraping

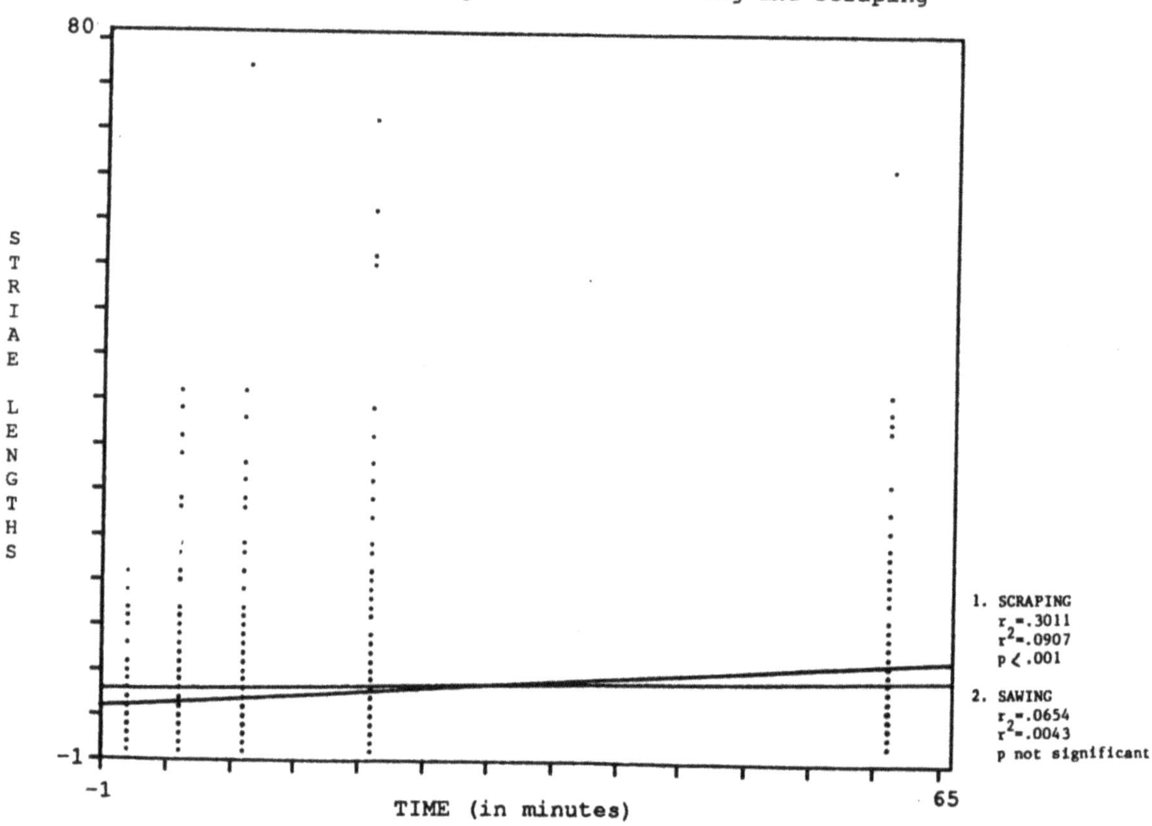

Fig. 24

Regression: Pits on Time for Wood Sawing and Scraping

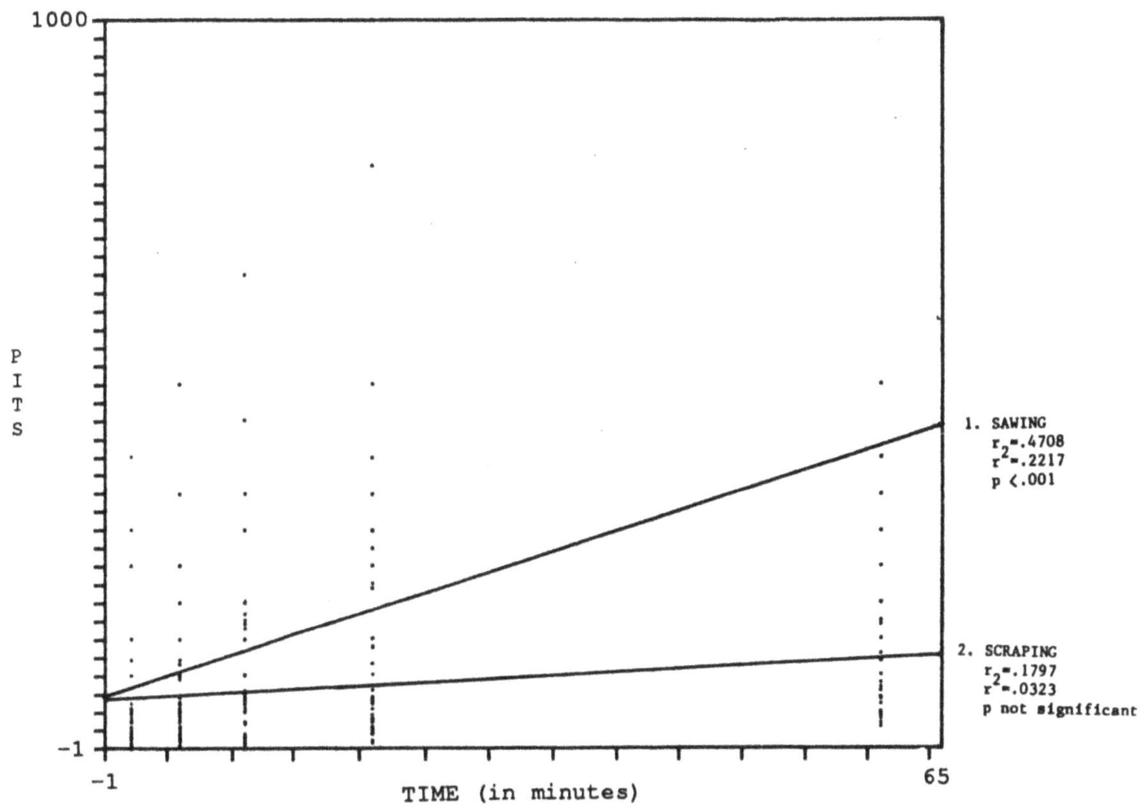

Fig. 25

Regression: Linear Grooves on Time for Wood Sawing and Scraping

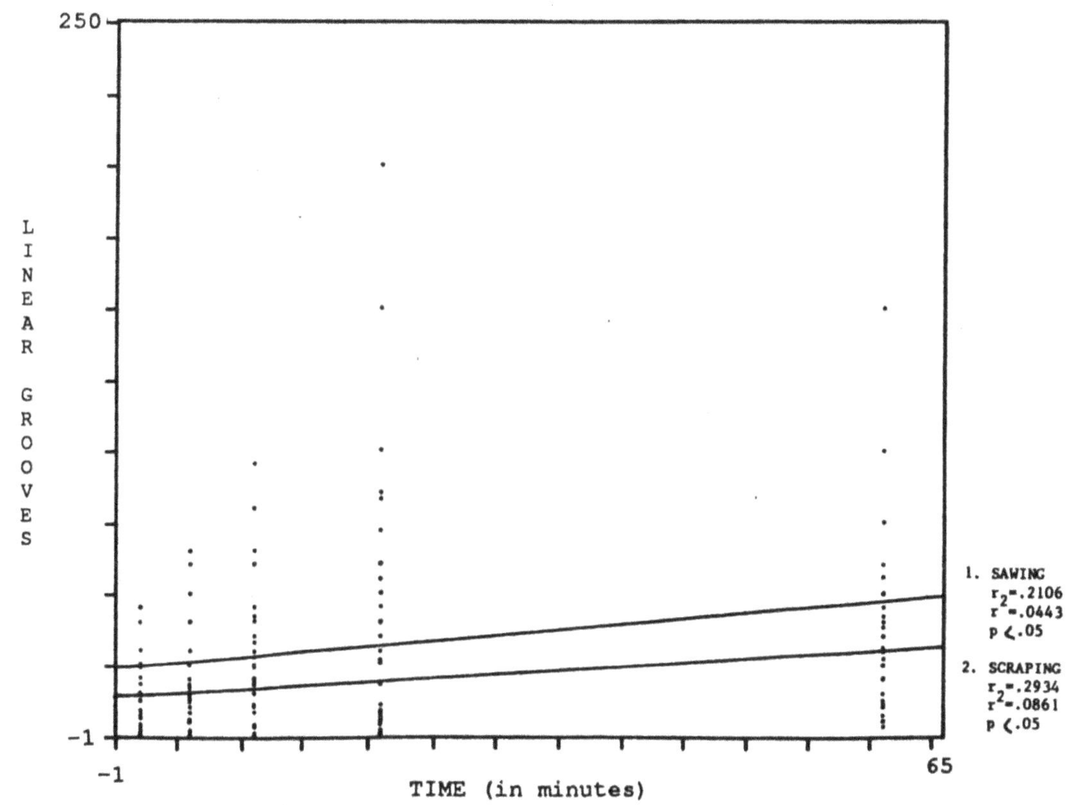

Fig. 26

Plate 41. SEM Photo of Quartz Edge after Sawing Oak
Note Pits and Linear Grooves 1600X

Plate 42. Quartz Edge After Sawing Wood (Yew)
Note Attrition of Surface Light Microscope 200X

Plate 43. Quartz Edge used for Sawing Antler, 10 minutes
Note Attrition of Surface and Striae
Light Microscope 200X

Plate 44. Quartz Edge after Sawing Wood (Eucalyptus), 60 minutes
Note Rough Surface Texture Light Microscope 200X

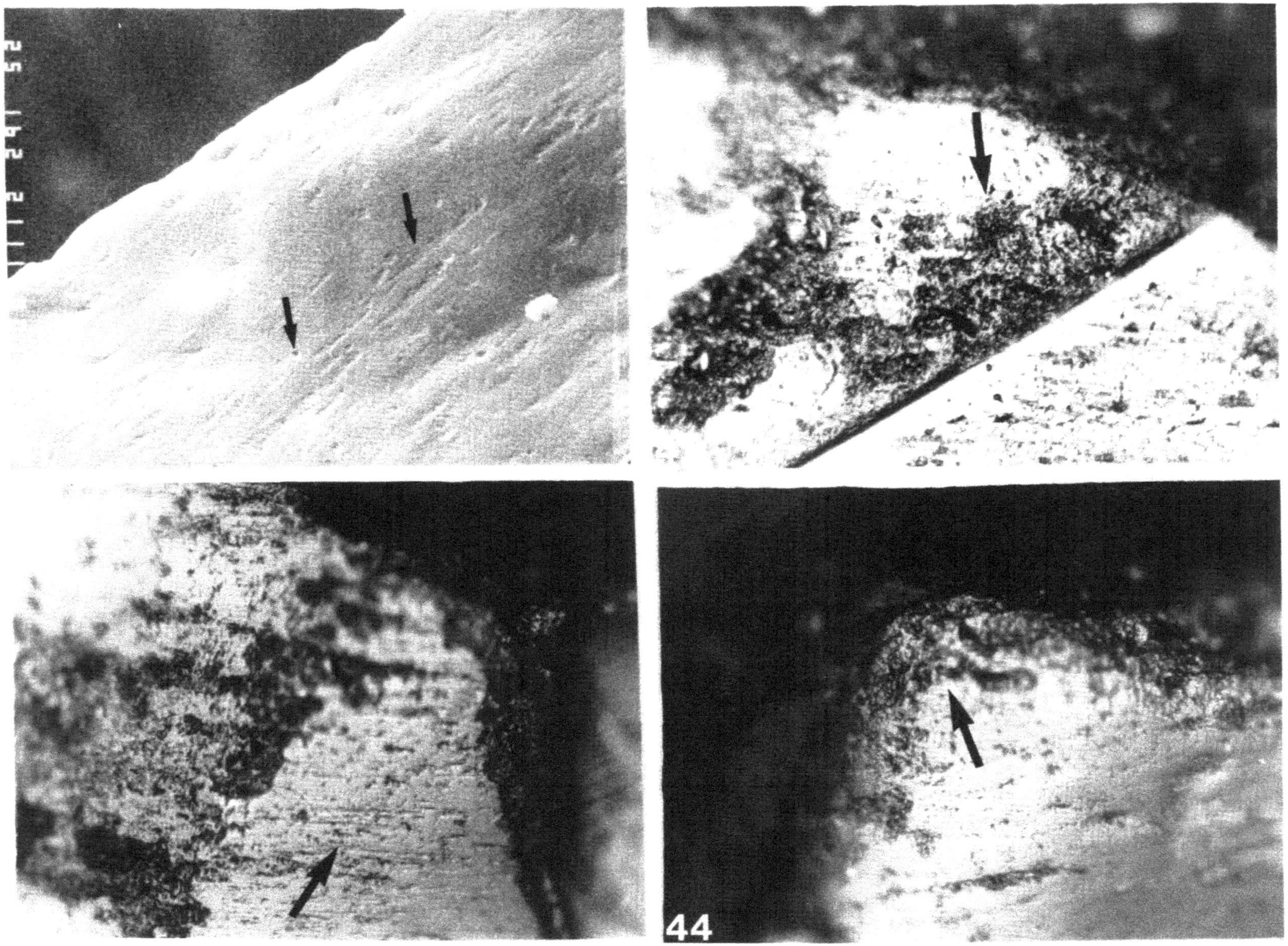
44

Plate 45. Quartz Edge after Scraping Hard Wood, 60 minutes
Light Microscope 200X

Plate 46. Quartz Edge After Sawing Antler, 60 minutes
Light Microscope 200X

Plate 47. Quartz Edge After Scraping Antler, 60 minutes
Note this piece has a smoother surface than
Plate 46 and more edge rounding Light Microscope 200X

Plate 48. Quartz Edge after Sawing Dry Antler
Note Nubbly Surface Light Microscope 200X

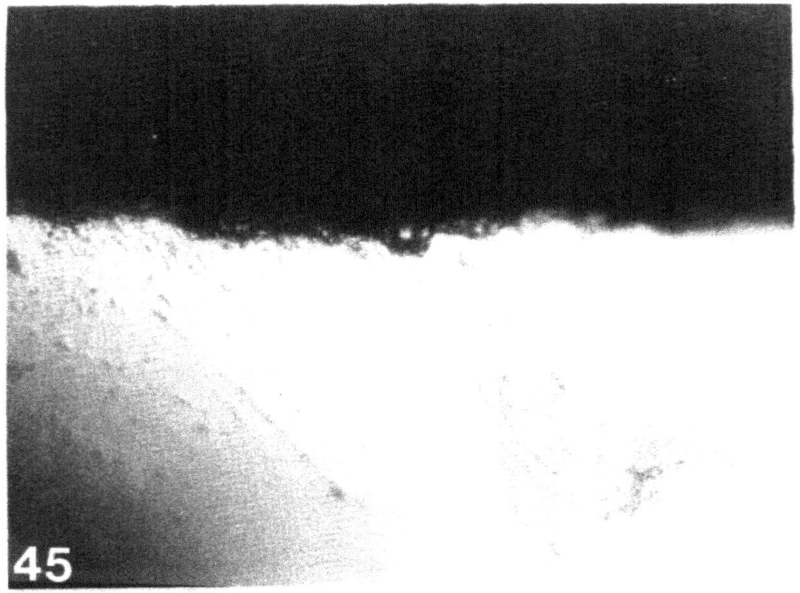

45

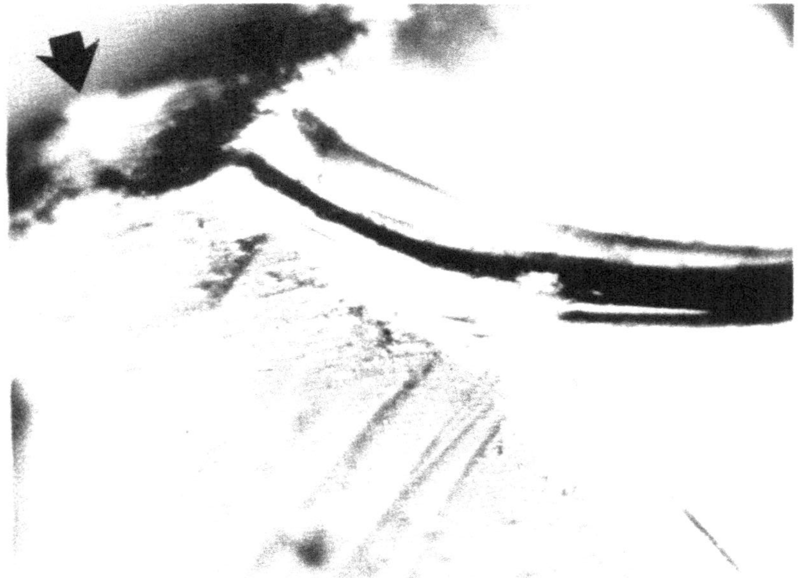

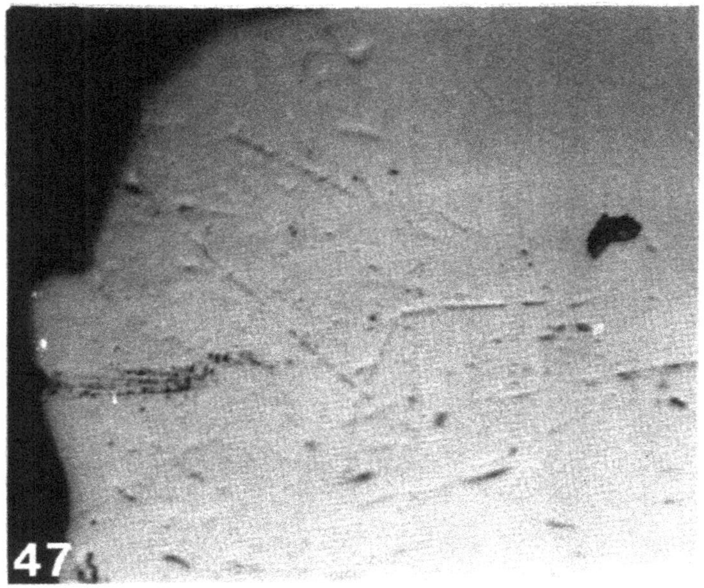

47

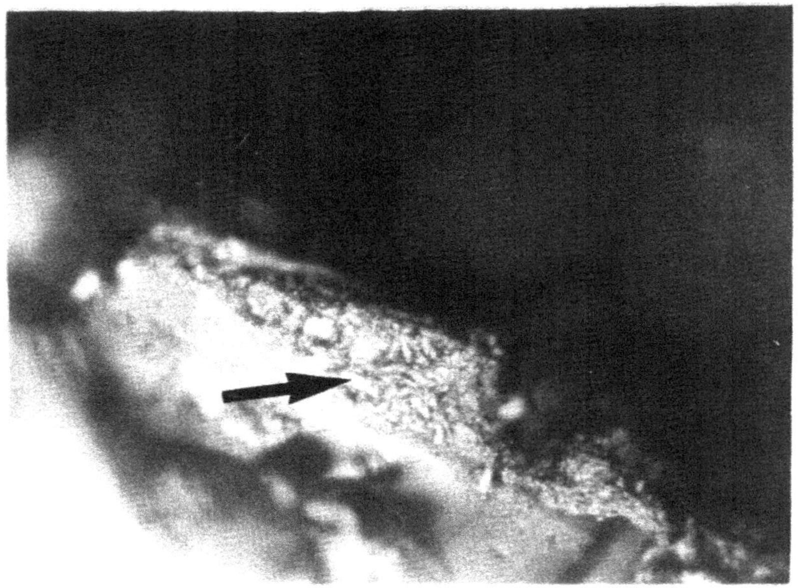

Chapter 6

Antler Working

Total Number of Flakes: 36

Total Number of Attribute Counts (5 per piece): 180

Number of Flakes used to Saw Soaked Antler: 14

Attribute Counts for Sawing Soaked Antler: 70

Number of Flakes used to Saw Dry Antler: 5

Attribute Counts for Sawing Dry Antler: 25

Number of Flakes used to Grave Soaked Antler: 13

Attribute Counts on Graving Antler: 65

Number of Flakes used to Scrape Soaked Antler: 9

Attribute Counts on Scraping Antler: 45

During the European Upper Paleolithic, working bone and antler was common, producing a variety of objects such as needles for sewing, or small figurines (Clark 1969; Piggott 1973). Antler working is not an activity found prehistorically in Africa due, undoubtedly, to a lack of fauna with antlers. I found, as did Keeley (1980:56), that working dry antler is an unproductive task, inflicting much damage on both the antler and the stone tool (be it quartz or flint), with little gain. Soaking the antler in water for just a few hours rendered it soft and more easily worked, and the stone tool edge remained sharp for a longer period of time. It is likely that prehistoric peoples worked antler while it was still fresh or soaked. There was no further advantage gained by soaking antler longer (24 hours to > 1 month) and, after about an hour of work, progress was aided if the antler was re-immersed in water for ten minutes. Unless specifically stated, all experimets described in this chapter were done using soaked antler.

<u>Scars</u> - Scalar scars were apparent soon after antler working began, followed by the appearance of step fractures. A combination of the two was visible after an hour of scraping or graving. Scars left on graver bits from use were much smaller than those seen on any other tools. Half moon breakages occurred intermittently but, only on the antler saws. This accords well with a typical edge damage on flint used in the same manner (Ibid:57-58).

<u>Pits</u> - Both scraping and sawing soaked antler produced copious amounts of pits on the tool surface, while graving soaked antler and, surprisingly enough, sawing dry antler, both produced on average, many fewer pits (Table 13; Plates 46-49).

Overall, both scraping or sawing soaked antler produced an

increase in the number of pits after ten minutes of use. This number more than halved itself after twenty minutes of sawing but, showed a corresponding increase in the number of comet-tailed pits present (see proceeding section and Table 13). The combined average of comet-tailed pits, however, is much smaller than that calculated for pits; but even so, after 96 minutes of sawing, comet-tailed pits far outnumber the others.

As mentioned, pitting reaches a peak after ten minutes of scraping but, oddly enough, there was no evidence of pits after scraping continued for twenty minutes. They appeared again on quartz used to scrape soaked antler for 60 minutes and 96 minutes. A regression analysis run on the number of pits and their relationship to the amount of time a tool was used, showed only scraping antler, and all antler work combined, produced a strong correlation with time (Fig 27). The pits seen on those quartz tools used for working soaked antler was absent from flint tools used in the same manner.

<u>Numbers of Pits</u>

	N	Mean	S.D.	Range
All Antler Working	180	93.55	170.33	0-999
Sawing Soaked Antler	70	125.68	168.38	6-999
Sawing Dry Antler	25	31.12	20.06	0-70
Graving Soaked Antler	65	30.49	39.62	0-200
Scraping Soaked Antler	45	134.66	246.26	0-999

N refers to attribute counts (5 per piece)

Table 13

<u>Comet-tailed Pits</u> - Although there are less than half the number of comet-tailed pits on all antler pieces combined compared to 'regular' pits (see Tables 13 & 14), a strong correlation with time is still expressed (Fig. 29). This type of pit first becomes noticeable after using a tool for five minutes on antler. Comet-tailed pits differ from 'regular' pits as they have shallow "tails" trailing after the heads of the pits and were found <u>only</u> on quartz tools used for working antler and bone. Why they are found only after working bone and antler remains, at present, a mystery. They are found on flint tools <u>only</u> after processing soft plant material and they are called "comet-shaped" pits; first described by Whitthoft (1967; Keeley 1980:60-61). None, however, of the quartz flakes used for cutting soft plant material showed any evidence of this type of pitting.

It is possible that they are caused by small pieces of bone or antler dragged across the surface of the polish (rather than the quartz surface itself) leaving a "tail" behind as the polish forms. Perhaps antler and bone polishes are not as 'fluid' as has been suggested for wet hide polish (Chapter 3) where few striae form. They suggested for wet hide polish (Chapter 3) where few striae form. They are extremely shallow features; documenting their presence with the scanning electron microscope required a tilt of approximately 30 degrees and 500X magnification to make them visible (Plate 9). Using the light microscope, however, they are easily seen at 200X (Plate 15).

The greatest number of these comet-tailed pits was found on those tools used for sawing, with the second largest quantity found on gravers and finally, the smallest number of comet-tailed pits was found on those quartz tools used for scraping. As with pitting, comet-tailed pits reach their greatest numbers after ten minutes of sawing and then peak again after sawing antler for 96 minutes. A significant regression was achieved when comet-tailed pits were correlated with time while sawing antler.

In general, the mean number of comet-tailed pits increased consistently during graving activity, and produced a strong regression when correlated with time (Fig. 29), far surpassing the number of 'regular' pits found on the same tool (Appendix 3). Scraping produced very few comet-tailed pits - many fewer in fact, than 'regular' pits. There were no comet-tailed pits seen on those tools used for either twenty minutes or ninety-six minutes of scraping antler but, inexplicably, a large increase in their numbers was found on the quartz scraper used for sixty minutes.

As with striae, the orientation of comet-tailed pits can be used to infer action, or direction of use. They run parallel to the working edge of a tool used for sawing, and at right angles to the working edge on a tool used for either scraping or graving. The location of the "heads" of the pits is more variable when scraping or graving; they sometimes orient towards the working edge and sometimes away from it.

Numbers of Comet-tailed Pits

	N	Mean	S.D.	Range
All Antler Working	180	38.99	93.99	0-999
Sawing Soaked Antler	70	58.31	132.69	0-999
Sawing Dry Antler	25	NONE		
Graving Soaked Antler	65	34.6	62.27	0-300
Scraping Soaked Antler	45	15.28	27.70	0-100

N refers to attribute counts (5 per piece)

Table 14

Striae - Where present, striae were generally short, broad and shallow, except on tools used for sawing, where the majority of the striae were long and narrow. This contrasts nicely with Keeley's description of striae he found on flint tools used for working soaked antler; most were were short and narrow (Keeley 1980:56). It should be noted that relatively few striae are to be found on either quartz or flint tools used for working antler (Ibid). When all antler working is combined and statistically examined, a significant (though not terribly strong) correlation emerges between the number of striae and the amount of time a tool was used ($p<.05$). The correlation loses its significance, however, when regressions are run on each activity separately (Figs. 30 & 37). Both sawing and scraping produced approximately the same number of striae, (approx. 4.1), while graving produced many less.

The longest striae overall, were those seen after sawing antler, while the shortest striae were those found after graving (Table 16). Only graving and scraping antler produced a significant correlation between the amount of time the tool was used and the lengths of the striae produced (Fig. 31). Hence, the longer scraping or graving continued using the same tool, the longer the striae produced.

Numbers of Striae

	N	Mean	S.D.	Range
All Antler Working	180	3.29	4.72	0-30
Soaked Antler Sawing	70	4.1	4.56	0-25
Dry Antler Sawing	25	3.0	2.22	0-7
Soaked Antler Graving	65	1.90	3.20	0-20
Soaked Antler Scraping	45	4.2	6.83	0-30

N refers to attribute counts (5 per piece)

Table 15

Striae Lengths

	N	Mean	S.D.	Range
All Antler Working	858	7.65	7.25	0-100
Sawing Soaked Antler	545	7.97	8.02	0-100
Sawing Dry Antler	63	7.47	5.57	0-30
Graving Soaked Antler	155	6.80	5.72	0-25
Scraping Soaked Antler	95	7.30	5.46	1-29

N refers to number of striae measured

Table 16

Linear Grooves - Sawing antler produced, by far, the greatest number of linear grooves; more than twice that produced by either scraping or graving (Table 17). Linear grooves became apparent after one minute of use regardless of the action but, only scraping antler produced a statistically significant correlation between the number of linear grooves and the amount of time a tool was used (Fig 28). As with striae, linear grooves followed the direction of use, making action recognition relatively straightforward.

There was a striking similarity between a tool used for sawing antler for ten minutes and a tool used for sawing yew for the same amount of time (Plates 42 & 43). The similarity between the two pieces is based on the abundance of linear grooves and it is reasonable to assume that one could easily be mistaken for the other in a blind test or an archaeological situation. This strong degree of resemblance was most prominent between antler and yew only, with other wood and other materials (such as hide or bone) being more distinctive. Beyond ten minutes of use, their attributes diverged, and a distinctive antler polish developed.

Numbers of Linear Grooves

	N	Mean	S.D.	Range
All Antler Working	180	20.98	44.01	0-400
Soaked Antler Sawing	70	36.74	64.11	0-400
Dry Antler Sawing	25	7.32	5.55	0-20
Soaked Antler Graving	65	8.35	17.12	0-120
Soaked Antler Scraping	45	14.73	16.54	0-80

N refers to attribute counts (5 per piece)

Table 17

Ridge Attrition and Ridge Rounding - The presence of attrition on the ridges and high points of the quartz microtopography was noted on all the tools; after one minute of sawing and graving, and after five minutes of scraping soaked antler.

Rounding of the high points first became visible after ten minutes of sawing, but no rounding was present on the scrapers, and rounding was found on gravers after an hour's effort.

Edge Attrition and Edge Rounding - Attrition along the working edges was evident on all the tools after one minute of use. Sawing antler for ten minutes produced rounding along the edge that was previously roughened by attrition. There was some smoothing also evident at this time which continued to develop and become more prominent with increased use.

After one hour, edge rounding created by scraping antler was not as smooth as that created after scraping bone for one hour (Plates 47 & 56). Attrition and abrasion were gradually obscured by smoothing and rounding only after twenty minutes of continuous graving.

Polish and Surface Texture - Antler polish was bright, metallic and luminescent for all activities but, usage time and type of activity influenced when these polish qualities became distinct and identifiable. When sawing soaked antler, the polish had a slightly metallic luster after five minutes of use which increased in brightness and "luminescence" as use time increased. Those tools used for graving acquired a polish that was metallic-looking and appeared distinctly additive after five to ten minutes of use and, after one hour of graving, the polish appeared brighter though no longer as metallic.

A rough surface texture was intermittently present on all of the tools during the first five to ten minutes of use. This contrasts with flint tools, where a rough surface texture was not time related but, rather, seen only after sawing soaked antler (Plate 50). Smoothing of the polished quartz surface increased directly as use time also increased but, became a truly prominent feature only on those quartz tools used for an hour to scrape or grave antler.

"Melted Snow" - By far, the most prominent feature which distinguished an antler polish from other polishes found on quartz was the "melted snow" quality which is so characteristic of smooth antler polish on flint (Keeley 1980:56). While difficult to describe objectively, it is probably best illustrated in Plate 49. The quality of this type of polish compares with the fluidity of a melting snowbank or, the smoothness of a heavy piece of satin draped over a mound. This effect was marginally visible after five minutes of use on antler but became a diagnostic feature after one hour of use, regardless of the type of activity performed.

From these experiments, it now appears that the development of an antler "polish" passes through an early stage, (after about ten minutes of use), where attrition and abrasion occur; a subsequent stabilization of the edge develops after approximately 20 minutes of use, which enables the polish to accumulate along with edge rounding, smoothing and "melted snow" (Fig. 27). This process apparently continues upto and including one hour of use. When tool use exceeds an hour, the polish appears to be literally worn away, which again results in an increase in surface attrition and pitting on the tool. This process was was visible on two tools used incrementally; one for scraping antler for a total of ninety-six minutes (1',5',10',20, and 60' increments), and the other for sawing antler for the ninety-six minutes. Both tools were cleaned and examined after each time segment was completed. After 96 minutes, there were numerous linear grooves and pits present on both tools along with a great deal of edge attrition - far greater than what was seen after sixty minutes (Appendix 3).

Rounding , associated with a "filled in" or additive aspect was found in the depressions and grooves but, the overwhelming feature was pronounced attrition which caused the polish to resemble a well developed wood polish. An antler and a wood polish are dissimilar in certain respects, with a "classic" wood polish being slightly rougher and lacking both the "melted snow" aspect and the "filled in" appearance to its pits or depressions. It is likely that if use continued for longer periods of time that the polish would again form when the edge stabilized once more.

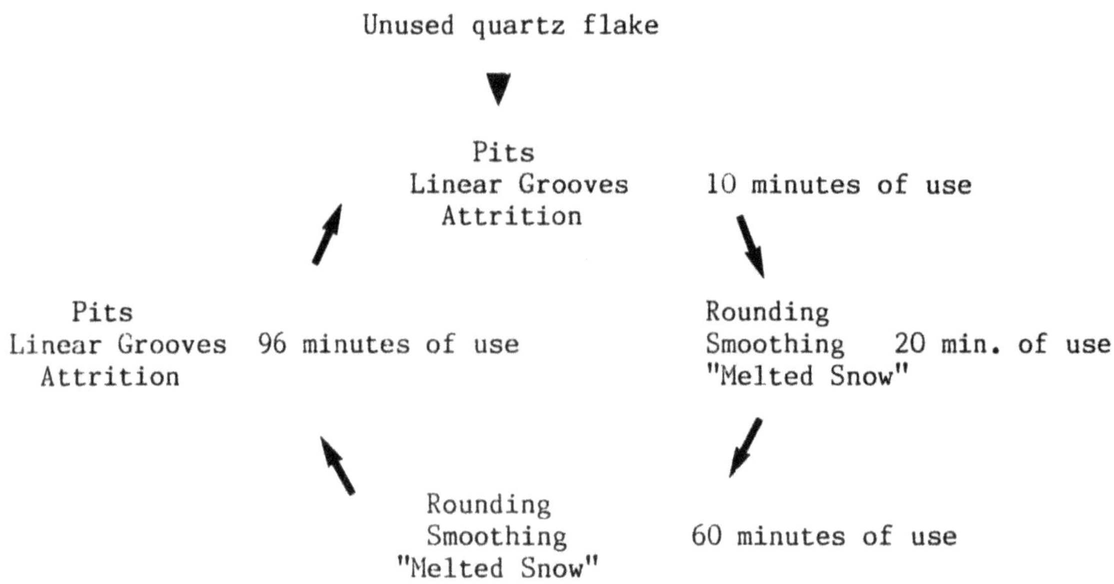

Fig. 27

Sawing Dry Antler - As stated previously, working dry antler was unproductive and rendered the stone tool, (flint or quartz), quite useless after a relatively short period of time while accomplishing little in the way of fashioning the antler. I did complete one "set" of sawing experiments on dry antler to determine if there was a visible polish or diagnostic damage on the quartz tools. The ridges and high points on the tool surface were more dramatically effected than those tools used on soaked antler. There was no rounding of high points noted at any time (Plates 48 & 51).

Edge attrition and severe edge damage in the form of scalar and step fractures, were the two overwhelming features of the quartz tools used for sawing dry antler, in fact, edge attrition became visible after only sixty seconds of use. There were very few attributes such as linear grooves and pits or striae present. In Figure 27, there is no regression line drawn for the relationship between number of pits and sawing dry antler. This is because the slope of the line was computed as zero, meaning that the number of pits bore no relation to the amount of time the tool was used. No comet-tailed pits were seen on any of the tools used on dry antler and, neither the number of linear grooves seen, or the number of striae present bore significant relationships to the amount of time the tool was used (Figs. 29 & 31).

Striae lengths on the dry antler saws were actually longer than those measured for both scraping and graving soaked antler but, were less numerous (Table 16). Narrow striae were found on those tools used for one minute and ten minutes but, broad, shallow striae were the only type observed on the five minute and sixty minute pieces.

Sawing dry antler with quartz never produced any discernable polish or "melted snow" effect but, some polish was noted on the flint tools used for sawing dry antler; being bright, reflective, rough, and discontinuous down the worked edges. The polish never developed as well as the polish on flint saws used for working soaked antler but, it accumulated in a similar manner; collecting on high points and edge protrusions as well as developing along the perimeters of the damage scars. There were a few narrow striae noted on the surface of the flint tools which ran parallel to the worked edge. Edge damage on flint was severe, comprising numerous of half moon, scalar and step fractures.

Summary - In summary, polish produced from working soaked antler using quartz tools had two distinguishing characteristics: the presence of comet-tailed pits and the "melted snow" effect. While working bone using quartz tools also produced comet-tailed pits, the "melted snow" phenomenon (which may become visible after only five minutes of use), was never seen.

Working unsoaked or dry antler proved to be a fruitless task and I think it is safe to assume it was avoided whenever possible.

Regression: Pits on Time for All Antler Work

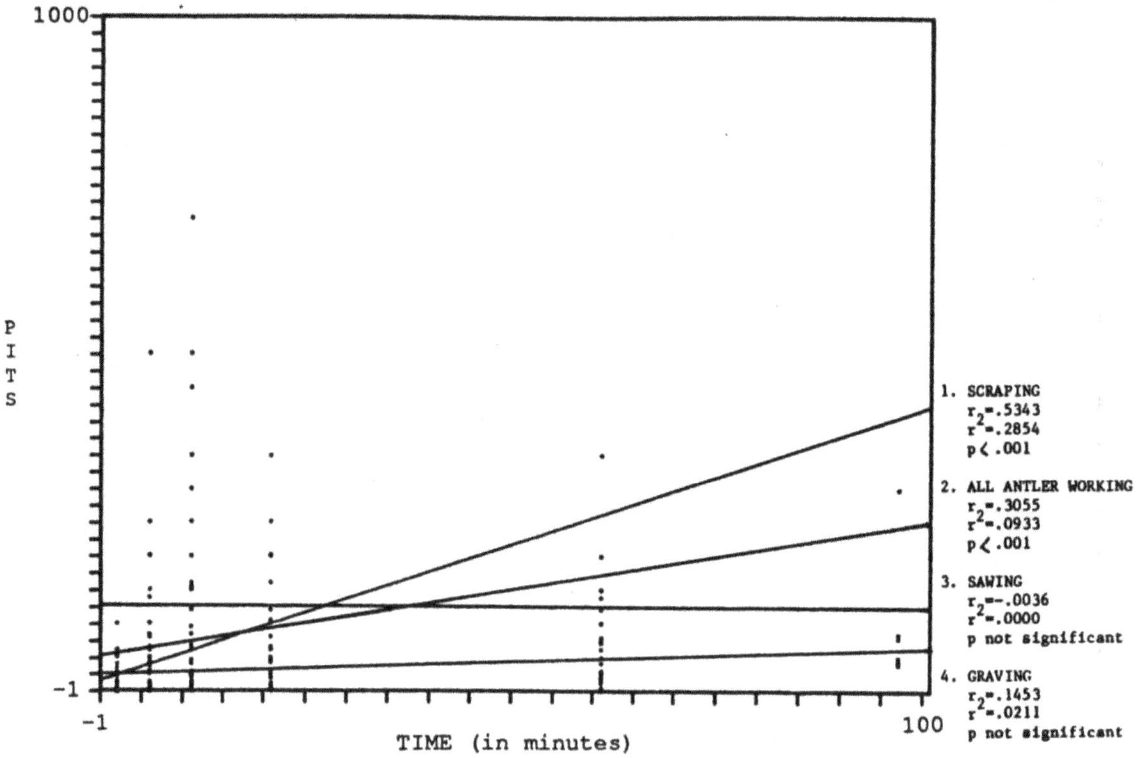

Fig. 28

Regression: Linear Grooves on Time for All Antler Work

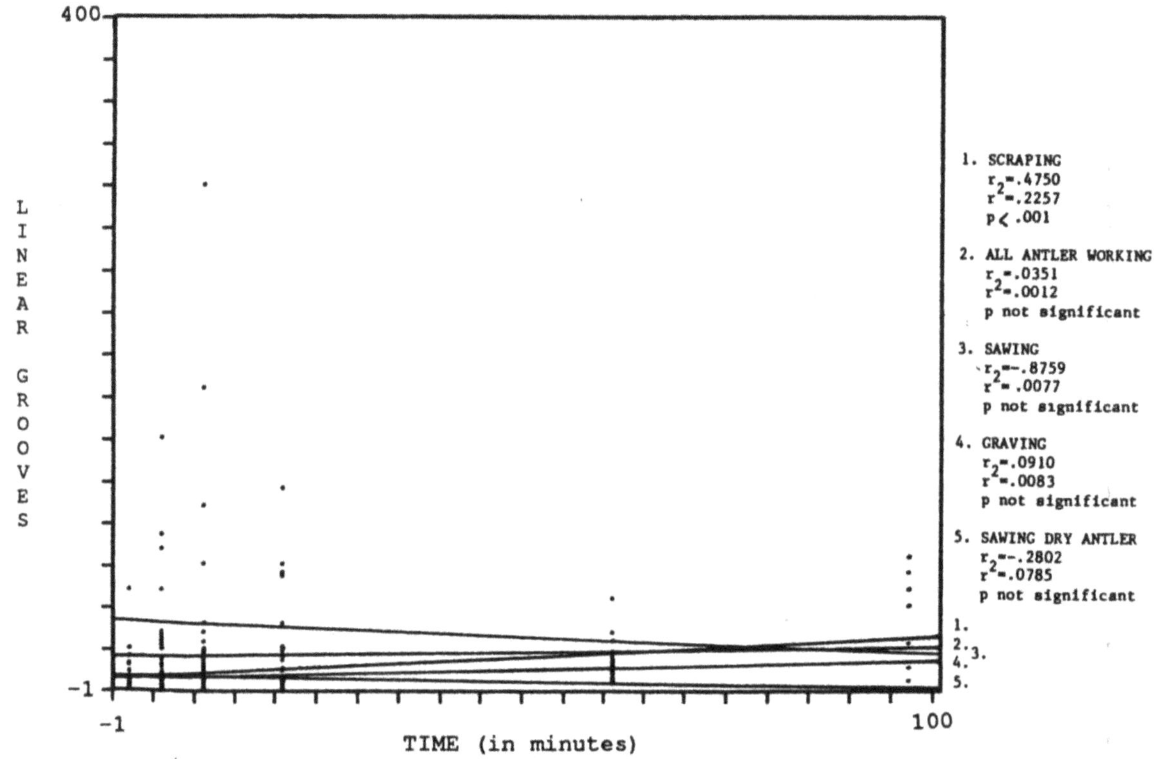

Fig. 29

Regression: Comet-Tailed Pits on Time for All Sosked Antler Work

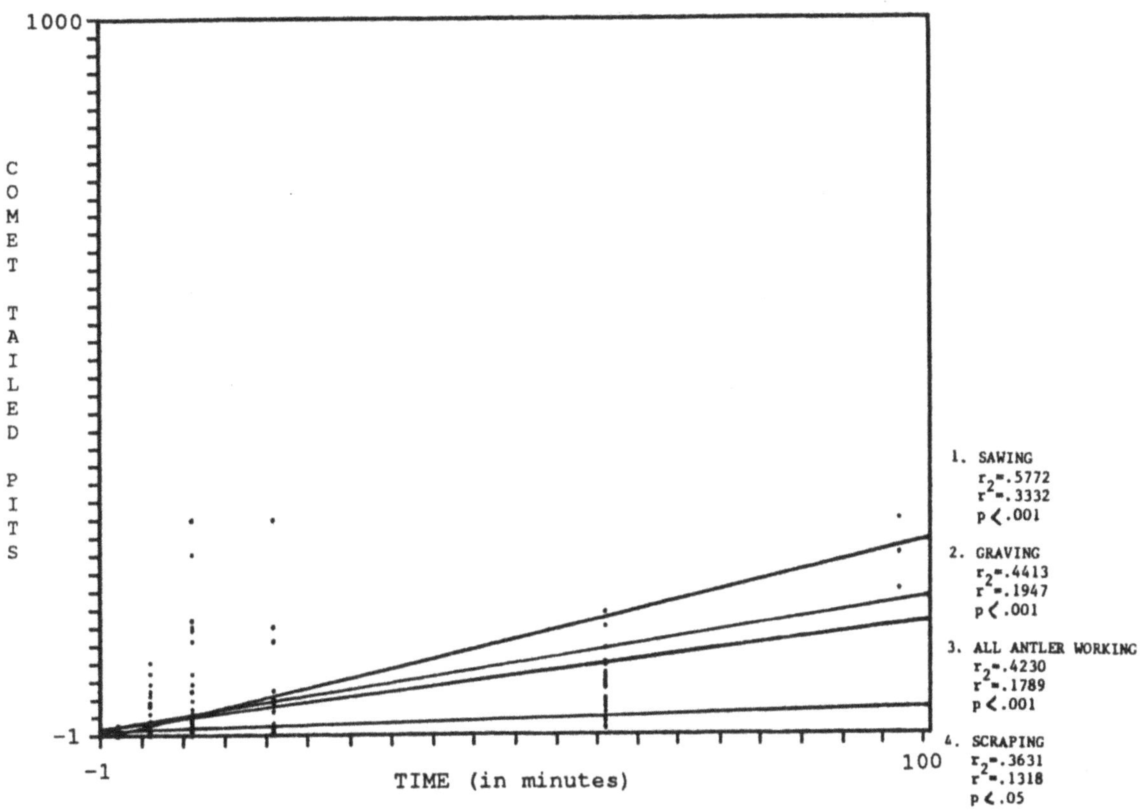

Fig. 30

Regression: Striae on Time for All Antler Work

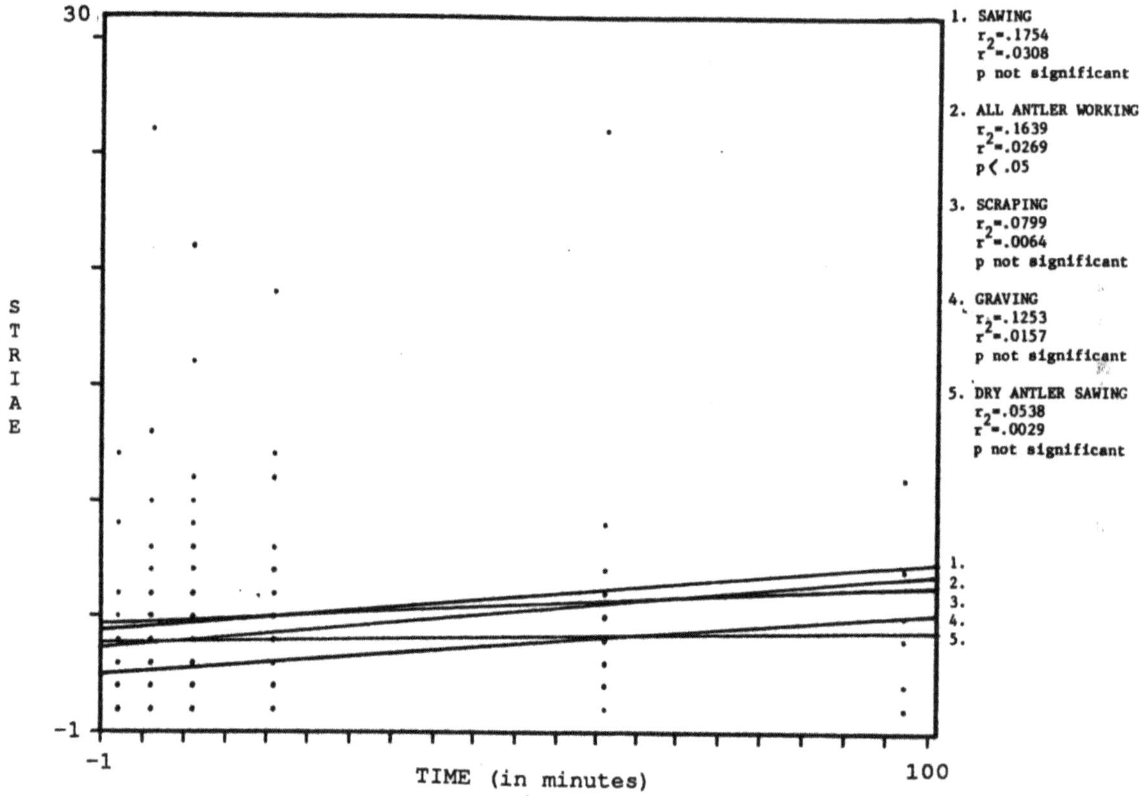

Fig. 31

Regression: Striae Lengths on Time for All Antler Working

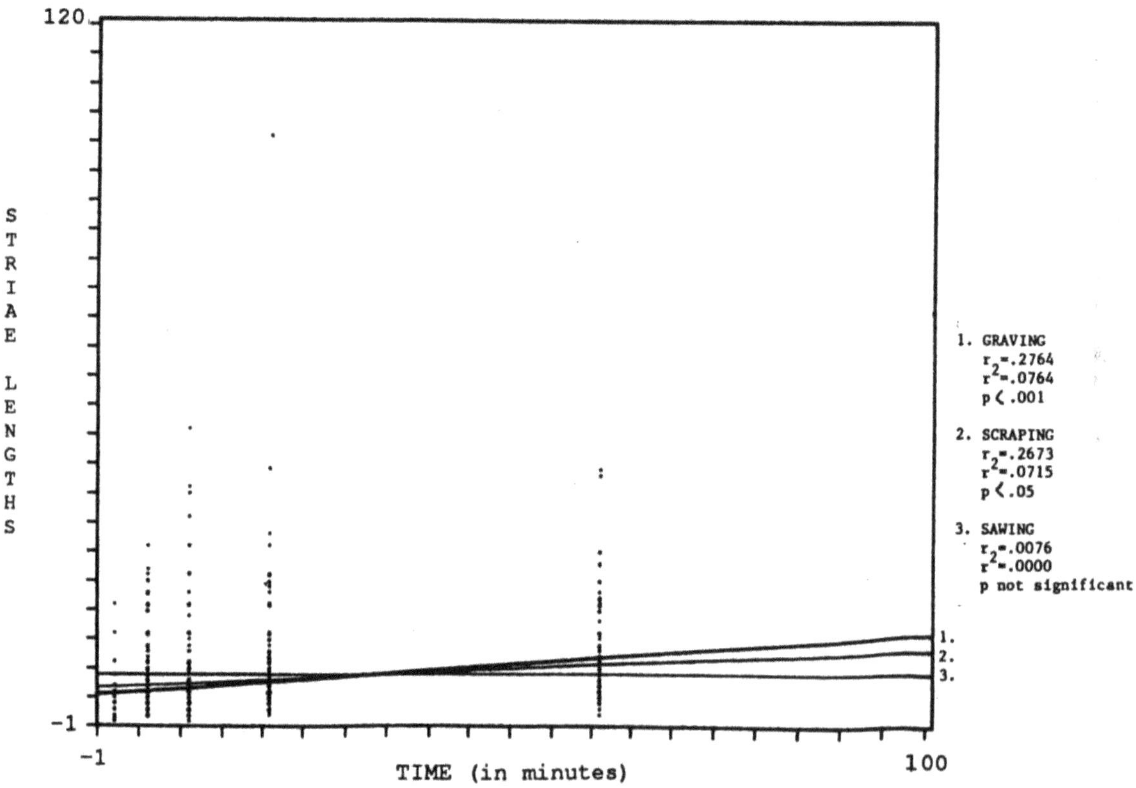

Fig. 32

Plate 49. Quartz Tip after Graving Antler
Note "Melted Snow" Aspect
Light Microscope 200X

Plate 50. Flint Edge after Sawing Antler
Light Microscope 200X

Plate 51. Flint Edge After Sawing Dry Antler
Light Microscope 200X

Plate 52. Quartz Edge after Graving Bone
Light Microscope 200X

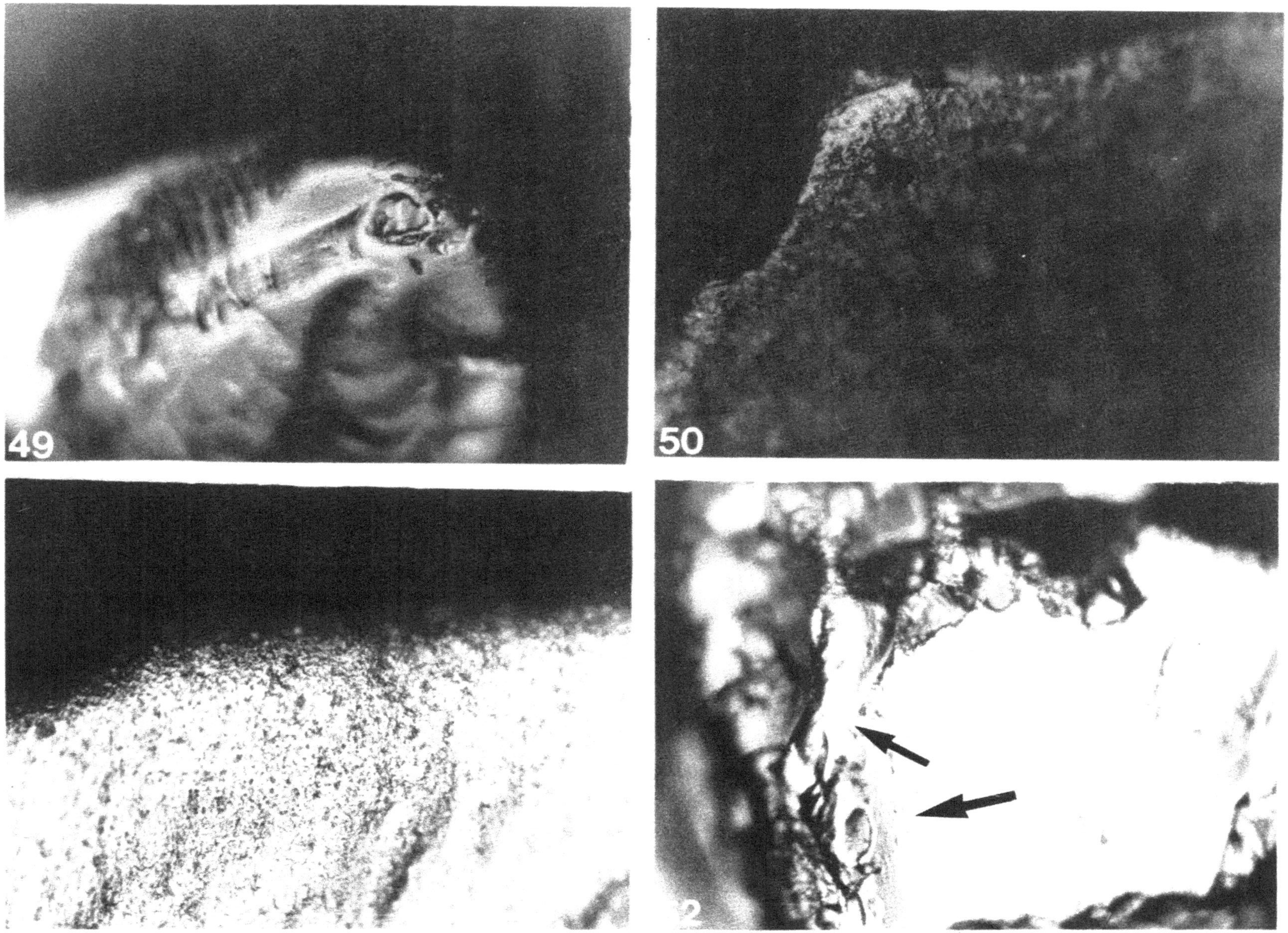
49
50

Chapter 7

Bone Working

Total Number of Flakes: 38

Total Number of Observations (5 per piece): 190

Flakes used for Sawing: 17

Flakes used for Scraping: 12

Flakes used for Graving: 9

The bone used for all these experiments was fresh cow bone. Working dry bone was unproductive when using flint tools and, as quartz is more friable than flint, working dry bone with quartz was not attempted. One piece of quartz used for sawing boiled lamb bone for fifteen minutes was examined and is discussed at the end of this chapter.

Scars - Edge damage from use was found along both faces of the working edges of the quartz tools used for sawing bone. The damage was comprised of predominantly scalar scars, with a few half moon scars as well. The presence of half moon scars seemed to be created by the acute edge angle of the individual tool used for sawing rather than being indicative of bone working itself. A larger sample size is necessary to confirm this. Half moon breakages are relatively uncommon also on flint used to saw bone (Keeley 1980:44). After one hour of use, step fractures were present on the quartz tools in addition to the scalar scars.

Scraping fresh bone resulted in discontinuous scalar scars and step fractures along the tool edges as soon as use commenced. These scars increased in number over time, until they were present continuously down the leading edge of the tool. Very little edge damage was sustained on the edge of the scraper contacting the bone.

Graving bone produced scalar scars, step scars and crushing on the graver bit; this crushing occurred as quickly as one minute after graving began. As with scraping, use damage occured on that edge opposite the side in contact with the substrate. Crushing, where present, obliterated all other scars.

Pits - The largest mean number of pits was produced by scraping bone when compared with either graving or sawing (Table 18). A regression run on the number of pits relative to the duration of use showed that both scraping and sawing correlated strongly with the amount of time a tool was used ($p<.001$) whereas, the number of pits produced by graving bore no relationship to time whatsoever (Fig. 33). It is interesting that, after twenty minutes of graving the number of pits reached its apex, only to decrease in number with longer periods of use (Appendix 3). Other attributes, such as polish formation or increasing stability of the edge, are either obscuring or preventing the formation of more pits.

Numbers of Pits

	N	Mean	S.D.	Range
All Bone Working	190	29.47	38.38	0-300
Sawing	85	43.0	49.25	0-300
Scraping	60	24.93	24.36	0-100
Graving	45	2.77	4.59	0-25

N refers to number of observations (5 per piece)

Table 18

Comet-tailed Pits - Graving bone averaged a larger mean number of comet-tailed pits than either scraping or sawing bone (Table 19) but, only sawing had a significant positive correlation between the number of comet-tailed pits and duration of use (Fig. 32). When all bone working was combined, a firm correlation between comet-tailed pits and duration of use was formed; this was more significant than the regression run on the occurrence of 'regular' pits. After scraping bone, the "heads" of the comet-tailed pits were generally oriented towards the working edge of the tool with their tails streaming behind them. While they did vary in the direction in which the "heads" faced (either towards the edge or away from it), the comet-tailed pits **always** parallelled the direction of use (Plate 56).

Number of Comet-Tailed Pits

	N	Mean	S.D.	Range
All Bone Working	190	24.65	43.52	0-300
Sawing	85	16.55	23.34	0-120
Scraping	60	25.6	39.28	0-200
Graving	45	38.71	67.73	0-300

N refers to number of observations (5 per piece)

Table 19

Striae - Striae were present on the quartz tools after only one minute of sawing or graving and after five minutes of scraping. There was no apparent statistical correlation between the number of striations produced and the amount of time a tool was used, regardless of the action (Fig. 35). Graving bone produced the largest mean number of striae overall, but sawing bone showed the greatest range in numbers of striae (Table 20 and Plates 52 & 53).

Only sawing produced a significant statistical correlation between length of striae and time ($p<.001$) even though scraping averaged the longest striae (Fig. 36). Relatively few striae formed while scraping or graving; perhaps they became obliterated by the rapid development of polish and comet-tailed pits.

The striae produced by working bone were broad and shallow as well as narrow and deep; the former being more prevalent when sawing, the latter when graving. This contrasts with bone polish on flint tools where narrow, deep striae were the most prevalent type (Keeley 1980:43). Striae on both quartz and flint tools tend to run parallel to the working edge in the case of sawing bone and at right angles to the used edge for scrapers and gravers.

Numbers of Striations

	N	Mean	S.D.	Range
All Bone Working	190	3.91	6.82	0-60
Sawing	85	6.04	9.03	0-60
Scraping	60	1.73	2.00	0-9
Graving	45	2.77	4.59	0-25

N refers to number of observations (5 per piece)

Table 20

Mean Striae Lengths for Boneworking

	N	Mean	S.D.	Range
All Bone Working	727	8.94	11.73	0-120
Sawing Bone	500	9.10	13.09	0-120
Scraping Bone	106	9.24	8.47	0-45
Graving Bone	121	8.01	7.37	0-36

N refers to number of observations (5 per piece)

Table 21

Linear Grooves - The gross numbers of linear grooves found on these quartz tools averaged approximately the same for all bone working and were a less prominent feature than either the comet-tailed or 'regular' pits. The only statistically significant correlation between linear grooves and duration of use came from sawing bone (Fig. 34). The prevalance of linear grooves on bone saws could be caused by micro-chips of quartz or bone becoming embedded within a track and continually dragged back and forth creating a groove, as the polish does not form fast enough to fill in and obliterate the grooves.

Working bone apparently does not drag micro-chips of quartz along its surface in the same manner as wood working, for example, which left a greater number of linear grooves on the tool. A quartz tool used for working bone, seemed to trap minute pieces of quartz and/or bone in the incipient polish as it formed, leaving a slightly roughened surface with what appeared to be microscopic fragments protruding from the polish.

Numbers of Linear Grooves

	N	Mean	S.D.	Range
All Bone Working	190	6.15	7.79	0-47
Sawing	85	8.35	9.45	0-47
Scraping	60	4.3	4.83	0-23
Graving	45	4.46	6.33	0-26

N refers to number of observations (5 per piece)

Table 22

Ridge Attrition and Ridge Rounding - There was little ridge attrition seen on the quartz tools used for graving bone; some was evident after five minutes of use but sporadically present thereafter. After twenty minutes of graving, attrition accompanied ridge rounding but, scraping bone produced no visible ridge attrition probably, because those flakes chosen as scrapers had relatively flat ventral surfaces. Rounding of ridges, however, was visible after sixty minutes of scraping and it is conceivable that any attrition present was obliterated by polish formation after an hour's use.

Bone saws suffered; the deterioration along the ridges was quite severe after an hour's use even though ridge rounding was present as well. Rounding and attrition appeared early on when sawing - after just five minutes of use.

Edge Attrition and Edge Rounding - The rounding of an edge after working bone equates with a smoothing of the working edge which blunted and thereby rounded it. This is quite different from the rounding described for dry hide polish which was created by severe edge attrition and produced instead, a rough, nubbly, uneven, blunted edge (compare Plates 55, 28 & 29).

Rounding of the edge became most prominent while graving fresh bone, even after short periods of time. Attrition of the working edge seemed to occur only during the initial "stages" of wear (one to five minutes) but rounding soon developed and became more pronounced as duration of use increased.

Edge rounding on bone scrapers was quite apparent after an hour's use and was caused by both attrition and smoothing. Attrition and smoothing were noted on all the bone scrapers, no matter how long they were used and both were also coincident on bone saws, with severity increasing over time. Attrition was present along the scraper edges soon after use commenced, and was visible after five minutes of sawing.

Surface Texture - There were two types of textural changes which occurred on quartz tools used for working bone: one was a smooth silky polish and the other was a slightly rough, sand-papery texture which appeared to come from microscopic chips of quartz and/or bone becoming caught in the developing polish (Plate 54). Some combination of the two was present on all tools used to work bone for twenty minutes or more. More equal proportions of the two textures were found on bone saws whereas those tools used for graving retained a much smoother polish.

Polish - Bone polish on flint retained the same attributes whether the activity was sawing, scraping, or graving (Plate 57). The polish concentrated on the high points or edge protrusions of the tool and, when well developed, became quite smooth. It is quite an aesthetic polish, being very bright and reflective, and often slightly greasy in luster. The polish was discontinuous down the working edges of both flint and quartz tools. As both bone and stone are unyielding materials, the polish formed on those areas of the microtopography which contacted one another. Bone polish on quartz formed quickly, becoming visible after five minutes of graving or scraping and after 10 minutes of sawing. Smoothing of the surface occurred first and then the bone polish became bright and metallic and, after graving, also became slightly greasy in luster. The polish looked additive although microprobe analysis and x-ray diffractive analysis revealed no organic residues left on the surface of the quartz.

Rounding and smoothing of the working edges were common features of a well developed bone polish (Plate 55) although, when not well developed, the polish was sometimes rough in texture. Striae were narrow and deep and followed the direction of use - striae created by sawing ran parallel to the working edge, and those left after scraping were at right angles to it.

After five minutes of use, pits, striae and depressions on the surface of the quartz flakes appeared to be "filled-in", as if something was being deposited or smoothed over them. If Anderson's (1980) hypothesis of polish formation on flint is correct and applicable to quartz as well as flint, then what is visible are the effects of the dissolution of the surface of the quartz from use and the formation of a silica gel on the working edges of the tools (see also Mansur-Franchomme, 1983). 'Silica gel' here refers to the results of the microprobe analysis - only silica was present on both the polished and unpolished area of the quartz. Mansur-Franchomme (1983), used X-ray analysis on experimental hide working tools, however, and found organics present after cleaning. She felt the organics must have been covered by silica because they "survived" treatment with hot H_2O_2. Clearly, this is one of the areas where further research can only aid in our understanding of the processes involved.

After an hour of sawing, a noticeable increase in edge attrition and edge damage occurred which removed much of the polish.

The polish formed after sawing boiled bone for fifteen minutes resembled more a wood polish than a bone polish. It lacked any comet-tailed pits which are so characteristic of working fresh bone or soaked antler, it was bright, discontinuous and not well developed overall but, concentrated on the higher points of the microtopography with attrition present along the edge. The lack of comet-tailed pits may be related to fat being drawn out by boiling and changes in the moisture content of the bone itself.

Cracks in the Polish - These are cracks which appeared in the surface of bone polish on flint tools as well as on quartz tools, although I found them to be much more numerous on quartz. Other researchers have seen this phenomenon as well (J. Dumont, L. Keeley, pers.comms., E. Moss 1983:92). The cracks run irregularly through polished areas as soon as the polish develops (Plate 56). This is a common phenomemon, which occured only when working bone and is thought, by at least one other microwear analyst (L.K.), to be caused by heat generated through the contact of the bone and the stone tool (flint in his case, quartz in mine). This suggests that the composition of a bone polish differs slightly from other polishes, perhaps in moisture and chemical (fat) content. Cross-sections of tools having well developed bone polishes could be compared with cross sections of tools having well developed silica polishes to see if any similarities exist. Flint would be the material of choice for cross-sectioning because it can be pulled apart easily with less fear of shattering than quartz (Meeks et al. 1982).

Summary - Bone polish forms very quickly and is characterized by distinct cracks running through the polish. Comet-tailed pits are present as well, paralleling the direction of use. The polish itself may be either very smooth and bright, often having a greasy luster or may be rough, with what appears to be tiny chips of bone or quartz 'caught' in the polish. It is possible to find both rough and smooth polish occurring on the same tool. Broad and shallow striae are present on bone working tools as are narrow and deep striae; the former being associated more often with sawing and the latter with graving.

Regression: Comet-Tailed Pits on Time for Bone Working

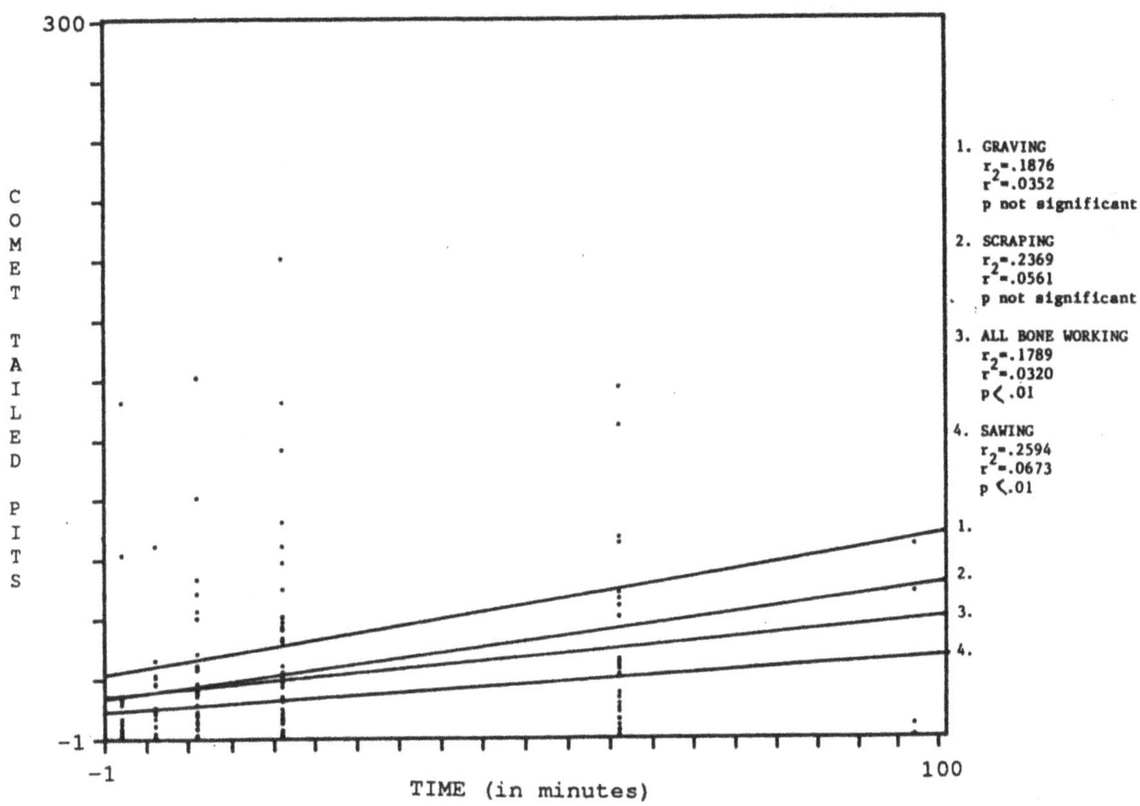

Fig. 33

Regression: Pits on Time for All Bone Working

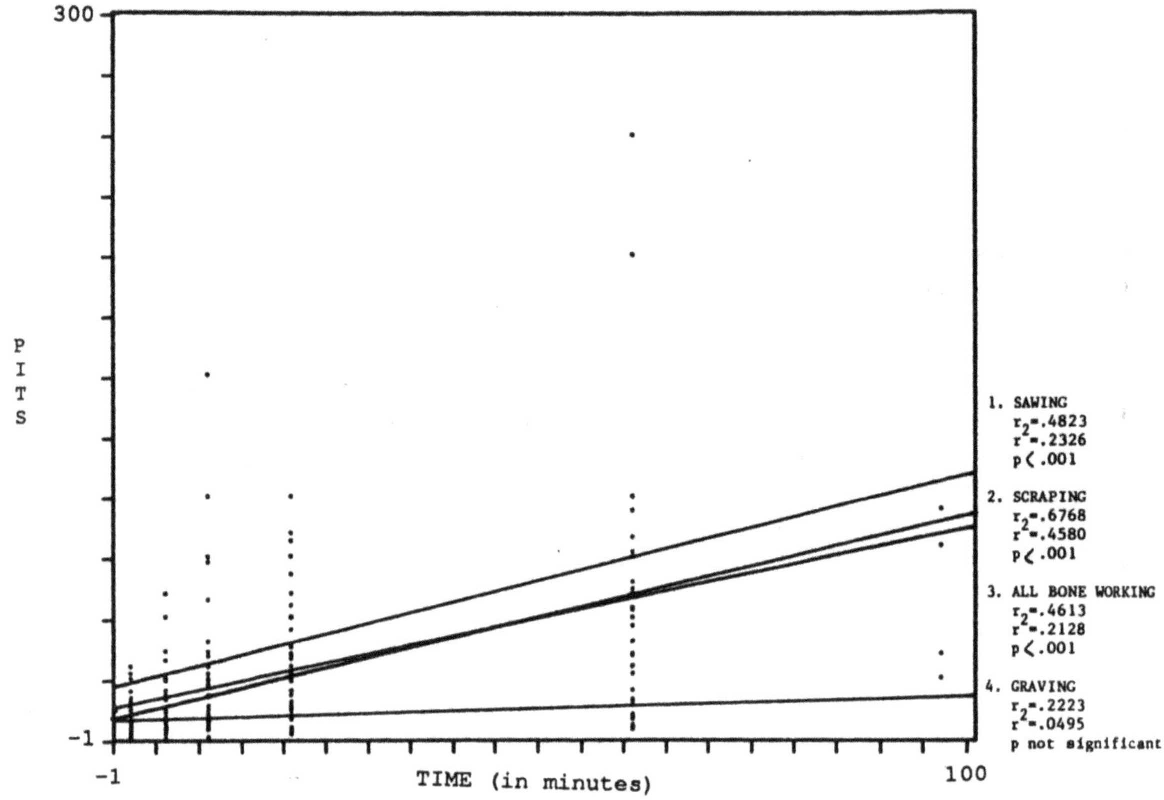

Fig. 34

Regression: Linear Grooves on Time for Bone Working

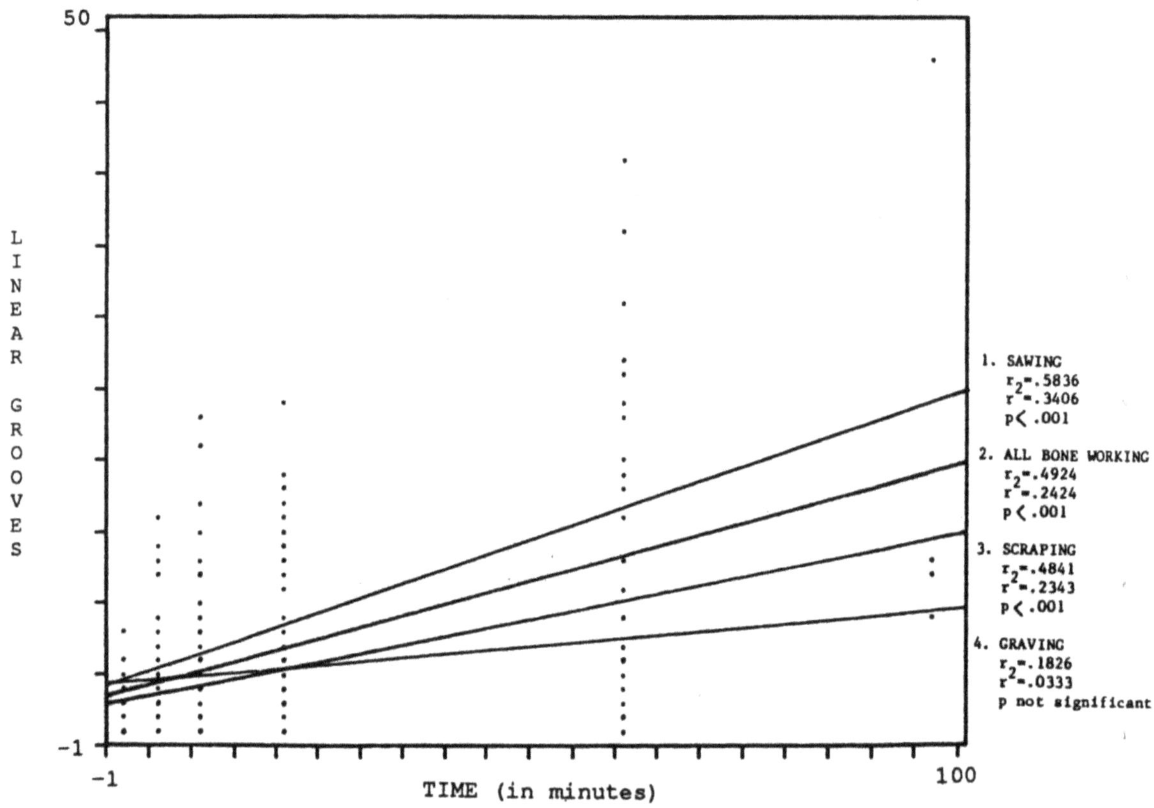

Fig. 35

Regression: Striae on Time for Bone Working (diff. actions)

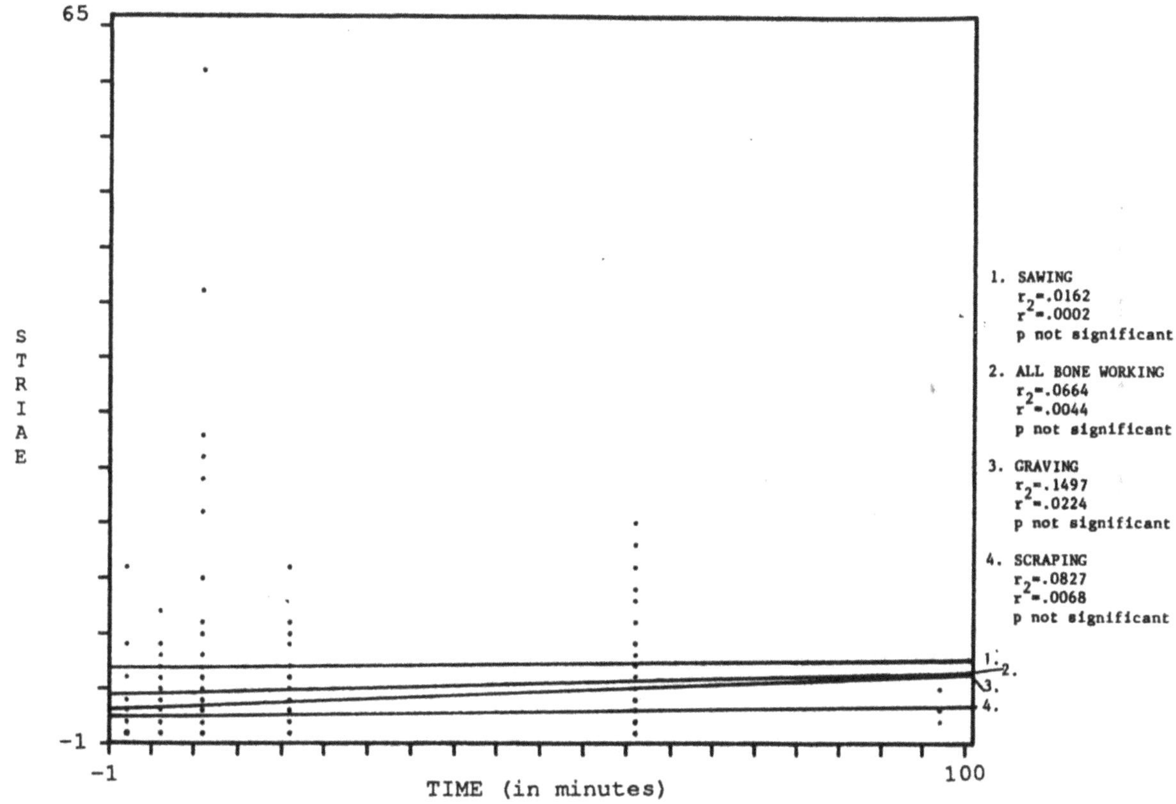

Fig. 36

Regression: Striae Lengths on Time for All Bone Working

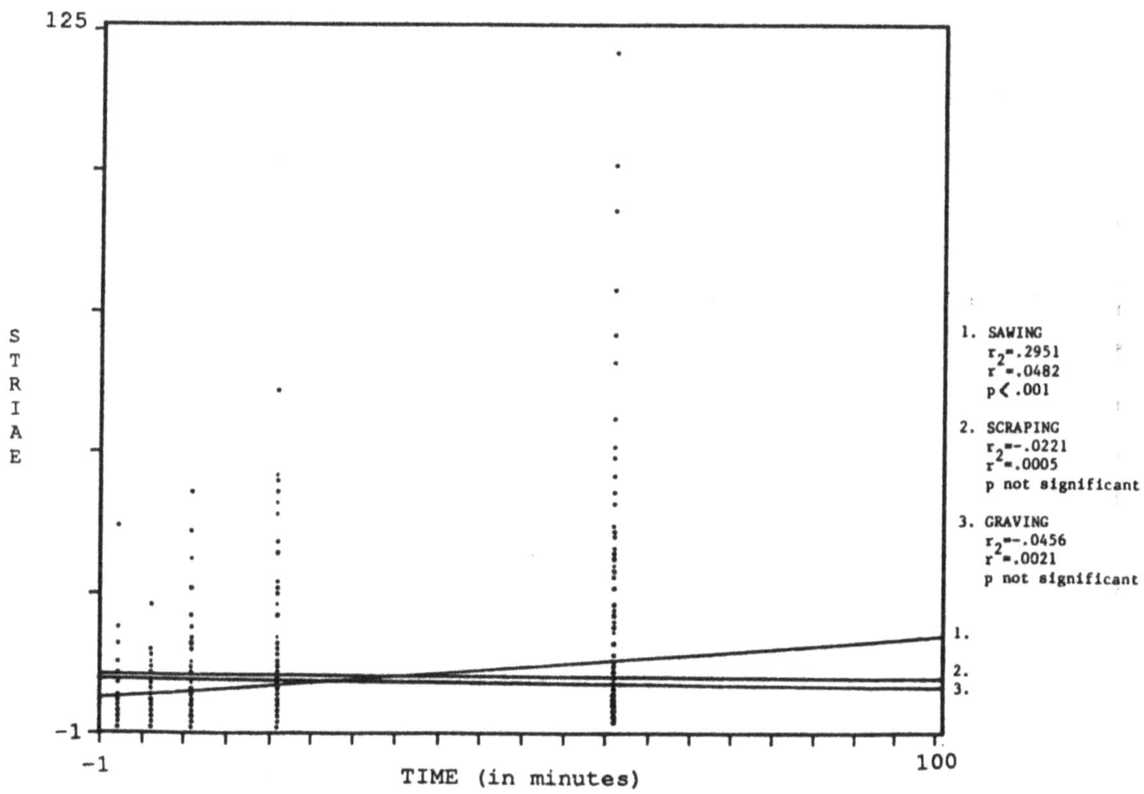

Fig. 37

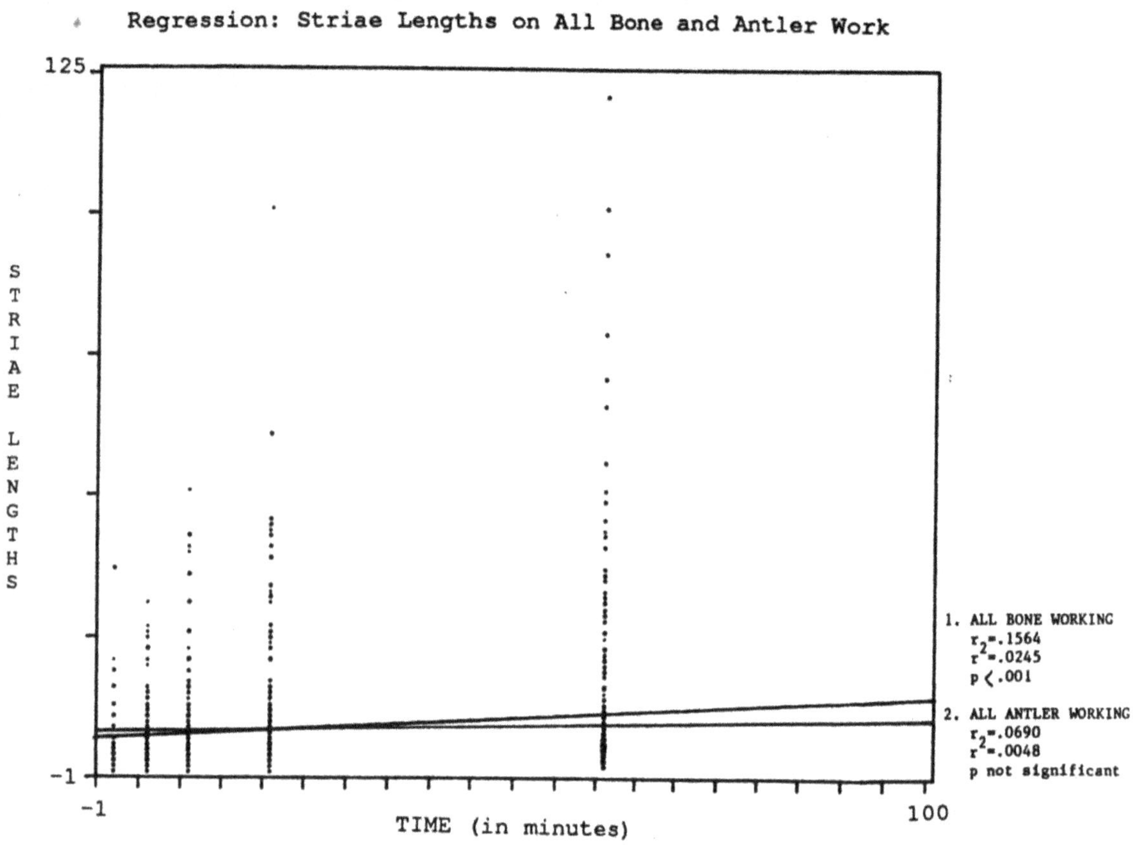
Regression: Striae Lengths on All Bone and Antler Work
125
STRIAE LENGTHS
-1
-1
TIME (in minutes)
100
1. ALL BONE WORKING
r=.1564
r^2=.0245
p $<$.001
2. ALL ANTLER WORKING
r=.0690
r^2=.0048
p not significant

Fig. 38

Plate 53. Quartz Surface After Sawing Bone
Note Striae and Compare with
Plate 55 Light Microscope 200X

Plate 54. Quartz Edge after Scraping Bone
Note Rough Surface Texture
Light Microscope 200X

Plate 55. Rounding and Smoothing of Quartz Surface
After Sawing Bone Light Microscope 200X

Plate 56. Quartz Edge after Scraping Bone, 60 minutes
Note Comet-tailed Pits and Cracks in Polish
Light Microscope 200X

53

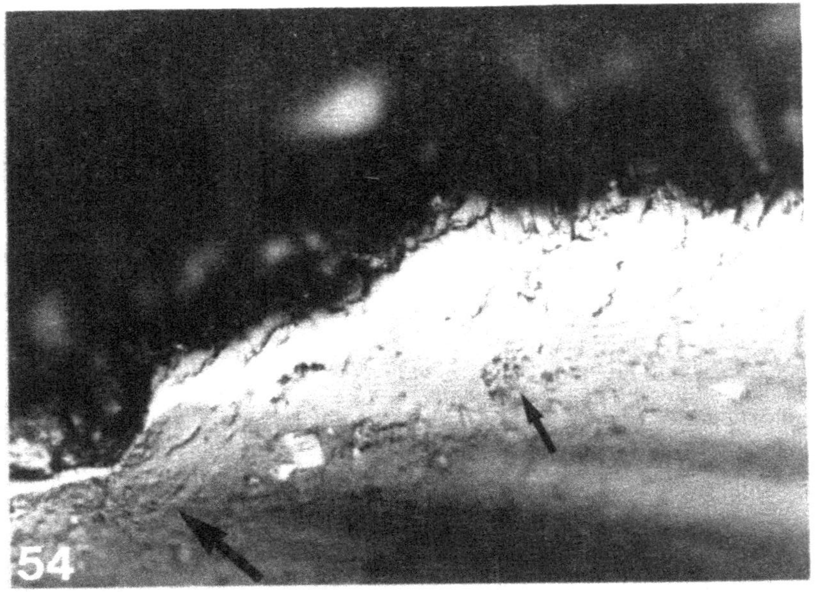
54

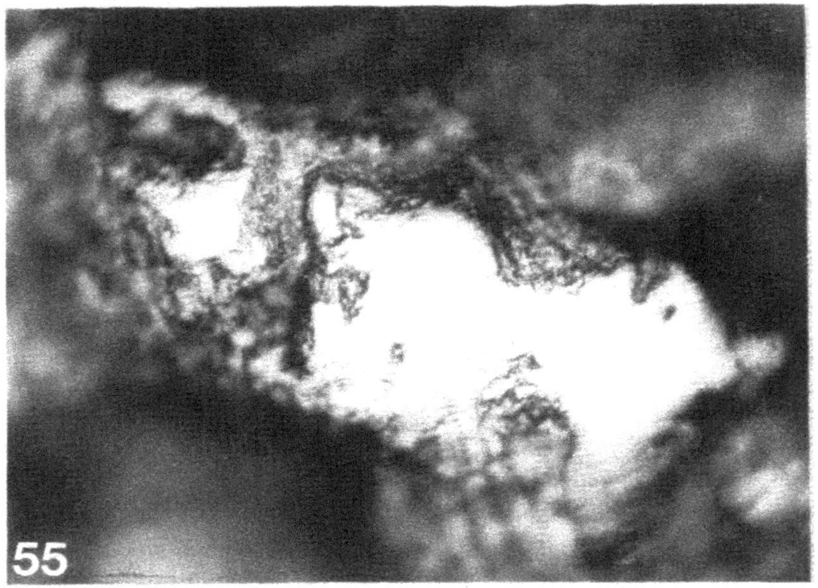
55

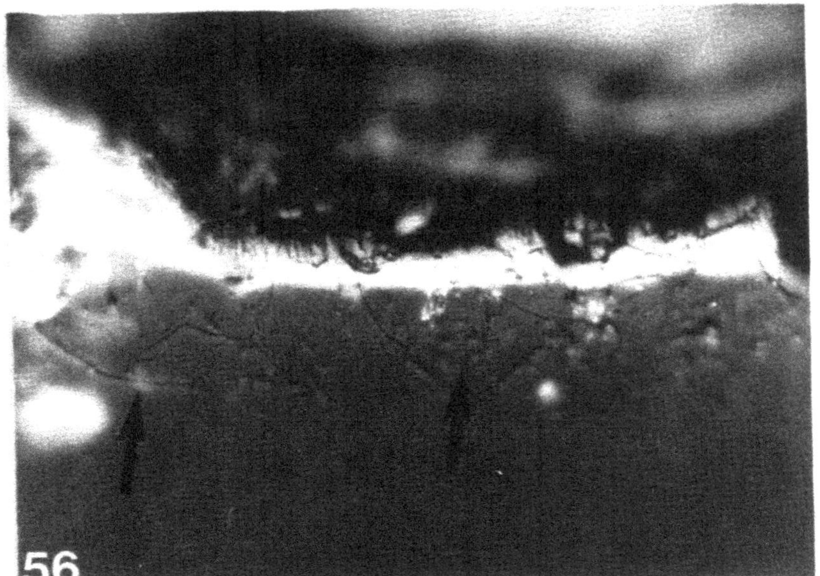
56

Plate 57. Flint Edge after Sawing Bone, 60 minutes
Light Microscope 200X

Plate 58. Flint Edge after Cutting Meat
Note Dull, Greasy Aspect
Light Microscope 200X

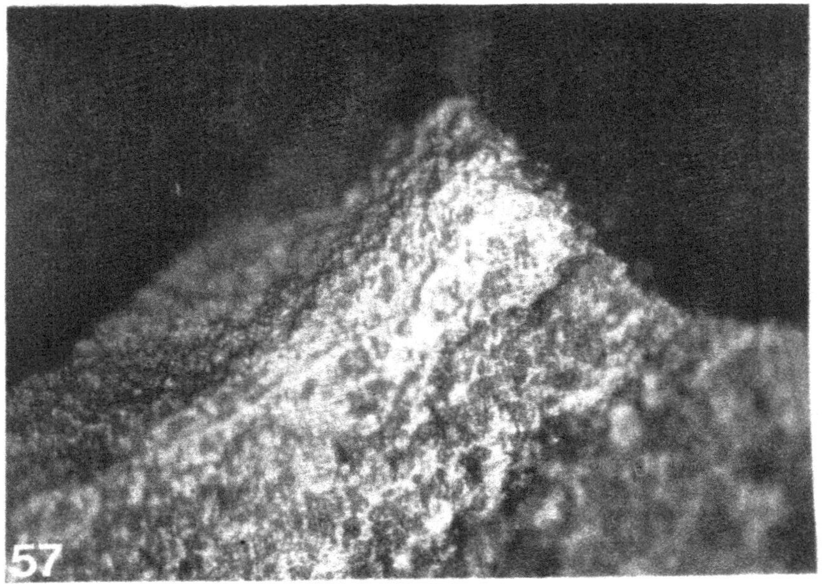
57

58

Chapter 8

Butchery

Total Number of Quartz Flakes: 5

Number of Attribute Counts: 0

One great advantage of using stone tools to early hominids is that sharp edged pieces of stone made it easier to acquire protein in manageable packages without the necessity of large, sharp canine teeth or claws. Scavenging and perhaps hunting of animals for meat was undoubtedly feasible and most probably practised by early hominids, using sharp-edged flakes and tools (see Blumenschine 1986; Bunn 1981; Bunn et al 1980; Jones 1980; Potts and Shipman 1981; Shipman and Rose 1983). There are numerous early sites in Africa (Clark and Haynes 1970;Isaac 1978; Leakey 1971), which are said to be butchery sites; where "artefacts are associated with the skeleton of a large mammal or with a group of smaller mammals" (Leakey 1971:258). Recent research on the bones from some of these sites show the existence of cutmarks which were made by stone tools. While this is strong evidence for butchery, there are those that argue that the spatial organization of many of these sites actually derive from the compounding of several episodes and agents, each contributing different materials (Binford 1981). The presence and identification of a meat or "butchery" polish (a combination of a meat and bone polish, for example, or evidence of cutting meat and fresh hide), on the stone tools found in association with the bones would allow archaeologists to directly link those tools with the bones.

The butchering of a fresh horse leg was accomplished using knives made from crystal quartz. Care was taken so that little contact with bone was made which might obscure or confuse the identification of a meat polish (e.g. no jointing was attempted in this experiment). All the knives cut through a portion of the hide, however, the majority of time was spent cutting meat away from the hide and and away from the ligaments.

The quartz knife used for an hour became extremely dull after twenty minutes of cutting and although the edge was wiped clean of adhering hair and other organic material, it remained dull though functional; with cutting capacity decreasing as time progressed. Undoubtedly, this was due to the accumulation of fat along the working edge which could not be wiped off, making the edge greasy and blunt. Louis Leakey used to suck the grease from the edges of his butchery tools when they became dull (J.D. Clark, pers. comm.) but, as the cause of death of this paticular horse was in doubt, that idea was seriously considered.

Recently, as part of another experiment, tools made of crystalline quartz were used to butcher four gazelle weighing between 140-160 pounds. The quartz tools worked extremely well and were very sharp. When flakes became blunted or broken, they were discarded. When these tools were examined microscopically for traces of wear, I was unable to find any discernable traces of polish using the light microscope or the SEM. The tools which were used to butcher the horse leg were used for much longer periods of time and therefore, if a meat polish was to form, it should have been visible on those.

Microscopically, however, none of the tools possessed distinguishing features which were unique to cutting and slicing meat. The surface of the quartz knife used for an hour on the meat was somewhat reflective and lumninescent under the light microscope at 200X, but in the absence of other diagnostic features such as pitting or linear grooves, this could not be attributed to use. A very slight rounding of the extreme working edge of the tool was noted. Edge damage too was very slight, with a few scalar scars present intermittently on both dorsal and ventral aspects of all the tools. Even these were not regular enough in their distribution to allow the tool to be labelled as "used" and just as easily could have been caused by dropping, trampling, or post-depositional movement.

Kjiel Knutsson, the Swedish microwear analyst working with quartz has been able to identify meat polish on his quartz tools. The polish he found was not very extensive, but wear features included broad mechanical striations, abrasion on ridges and patches of very rough polish. The wear is not diagnostic and can be very similar to the wear seen from working wet hide (K. Knuttson pers.comm.). At present, I have been unable to locate these features on any of my experimental butchery tools. The "mechanical striations" are most likely the result of grit in the meat or, perhaps, small fragments of bone or quartz which have been caught and scraped along the tool's surface. I, however, found what appears to be additive material on the surface of the experimental quartz butchery tools that were used for more than ten minutes (Plate 18). This feature has survived chemical cleaning and may be organic material adhering to the surface which was covered by silica, thereby enabling it to "survive" cleaning (refer to Mansur-Franchomme 1983). The ability to accurately assess the presence of butchery polish on quartz tools from some of the very early archaeological sites would be a valuable contribution indeed.

Meat polish on the experimental flint tools (Plate 58) was greasy, rough, dull to moderately bright, and present on both sides of the tools. Edge damage was indicative of cutting soft material and was slight; those scars that were present were small scalar scars and occasional small step fractures (Keeley 1980:53-55).

<u>Summary</u> - Unfortunately, at present, there are no visibly diagnostic traces which remain on quartz tools after cutting meat. The highly reflective and uneven surface of the quartz makes it difficult to discern subtle changes in "greasiness", "luster" etc. and as these are subjective terms anyway, the problem is compounded. It may be possible through further research and experimentation to isolate traces of polish by using dyes and markers which may pick up traces of certain chemical and organic residues. If there is indeed additive material being covered by silica, the potential exists to isolate and identify exactly what it is.

Chapter 9

Shell Drilling

Number of Quartz Flakes: 5

Number of Attribute Counts: 0

Polish left after working shell was found on experimental flint tools used to work shell by at least two researchers (N.Toth pers.comm and Moss 1983b). This polish was described by Toth as bright and irridescent with geometric "plates" resembling the bark on a ponderosa pine present on the surface of the tool (Fig.39). Moss, on the other hand, found that while boring shell "the seasonal growth rings of the shell often leave a distinctive track, like a ploughed field, clearly visible at >200X, which are accompanied by edge damage which leaves little polish," (1983b:151). The differences in their findings could be due to different species of shell being worked (Ibid).

Shell was used prehistorically and fashioned into pendents and beads, and was made into tools such as fishooks as well (Clark 1970; Flood 1983; Smith 1982). There is no reason to doubt that stone tools were used for fashioning the shell.

In an attempt to replicate this polish, both flint and quartz tools were used for drilling holes in abalone shell. All the borers endured tremendous amounts of crushing at the tips, (quartz more so than flint) and polish was not visible on either the quartz or the flint tools. The flint tools sustained numerous step fractures on their tips while crushing on the bits of the quartz drills was so severe that the damage scars were obscured and it was impossible to focus on the tip. Even without polish, in an archaeological context, the morphology of the tool as well as the presence of crushing and scarring on the tip would be a strong implication of use, although pinpointing the exact material the tool was used on would remain an unknown.

It is possible that these geometric "plates" are present on the borers but, were removed by chemical cleaning. A microscopic examination after cleaning in detergent but, before chemical cleaning of the flint borer used for one hour, showed some evidence of what may have been these "Ponderosa pine plates". After cleaning, however, I was unable to relocate them for photographing. They were not found on any of the used quartz drills before chemical cleaning.

Further experimentation, especially on shells such as ostrich eggshell would enlarge the sample size and perhaps produce significant results.

Chapter 10

Feather Working

Total Number of Quartz Flakes: 5

Total Number of Attribute Counts: 0

Feathers had many functional uses prehistorically, both in the Old and the New World; for ornamentation, clothing, hunting, bedding, etc.(see Pope 1918;1974 and Wheat 1967; inter alia). With this in mind, I shaved the quills of fresh pheasant feathers to document any damage and polish remaining on the quartz and flint tools after feather-working. Polish was noted on experimental flint tools used to work feathers by one researcher (G. Fennenga pers.comm. and unpubl.paper 1982) and I wished to replicate his results as "feather polish" has not been researched widely in the literature pertaining to microwear.

Fennenga found that polish left by feather working using flint tools was irridescent and had geometric plates similar to the "Ponderosa pine" plates documented on shell (Fig 39). I found no visible polish or "plates" on any of the experimental quartz tools and only a few scalar scars on the side of the tool opposite that contacting the quills. All the quartz tools used for feather working were made from milky white vein quartz which may have hampered identification of polish because of its irregular surface, large crystals and high degree of reflectivity.

The experimental flint tool used for one hour became serrated with half moon scars along the shaving edge and it had a moderately bright and slightly greasy polish present discontinuously down the contact edge. Small, bright areas of polish were found on the surface of the flint a small distance in from the tool's edge. Fennenga found geometric "ponderosa pine scales" were also visible some distance in from the working edge on his experimental tools. The presence of attributes some distance away fromn the working edge of the tool may result from the low angle at which the shaving tool is held during use, causing intermittent contact with the interior surface of the tool.

A small amount of polish was visible on the flint tool used for shaving feathers for twenty minutes but, this was not well formed. I found no "Ponderosa pine plates" on any of the experimental feather working tools, made of either quartz or flint. More experiments need to be carried out before definitive statements regarding the attributes feather polish can be made. The tools in question should be examined prior to chemical cleaning, perhaps rinsing only with water. If finger grease is a problem, then a rinse in detergent or acetone should be used.

Chapter 11

Summary and Conclusions

Perhaps the easiest way to summarize this work is briefly, in one sentence. Microscopic traces of use wear are present and identifiable on experimental stone tools made of quartz. To expand a bit, some activities leave traces of wear on the surface of the quartz tools which are more diagnostic and easily recognizable than others.

Working soaked antler, for example, left comet-tailed pits on the surface and a polish which has a unique fluid, "melted snow" like quality, similar to that seen on flint tools used to work antler. Striae were not common (mean = 3.3 ± 4.7) and, when present, were generally braod and short. There were twice the number of pits found on antler-working tools than comet-tailed pits (see Table 23) although, why this is the case remains unclear. More research is needed on the cause of comet-tailed pits. (Working fresh bone was the only other activity to produce these comet-tailed pits on quartz.) In general, an antler polish is bright and metallic; texture of the polish becoming smoother as the polish develops.

The two most diagnostic qualities distinguishing a bone polish from other types of use wear polish are the comet-tailed pits and the cracks which run through the polish. The polish itself concentrates along edge protrusions and high points, becoming bright and discontinuous down the edge of the tool. Of all the experimental materials worked, bone averaged the least number of pits overall (mean = 29.5 ± 38.4), and a relatively small average number of linear grooves (mean = 6.2 ± 7.8). It was not uncommon to find both a rough polish and a smooth polish on the same tool and it remains unclear as to why this occurs.

A soft plant polish was smooth and bright, similar to that described for flint tools also used to work soft plant material. The polish has a metallic sheen when developed, and extends into the low points of the microtopography in addition to concentrating on the high points. Pitting was always present, (mean = 56.6 ± 84.6) but striae, on the other hand, were never abundant (mean = $.84 \pm 1.7$). While "comet-shaped pits" have been noted on flint tools having silica gloss, I never found this to be the case on any of the experimental quartz tools used to work soft plant material. Plant polish and bone polish were the two most easily recognized and diagnostic polishes among the experiments using quartz.

Wood polish was more difficult to identify because it tended to develop inconsistently. When it was present, it was bright and metallic in luster, and sustained many more pits overall than any other type of polish (mean = 148.5 ± 217.4). A quartz tool used for working wood was more easily identified as such if polish developed in conjunction with pitting and linear grooves. I found an overlap between polish development in various activities; wood and antler sawing being the most obvious, and it was extremely difficult to distinguish between the two activities before ten minutes of use. After approximately ten minutes, the antler polish became distinct, and the attributes of the two polishes diverged. Care and familiarity with the raw material one is examining is essential, however, the fact remains that polishes <u>do</u> pass through stages of development where attributes overlap and, it is <u>not</u> always possible to confidently identify the material worked, although action is generally much easier to ascertain. This problem is further compounded, of course, when studying archaeological assemblages where other forces, such as soil acids, have been at work for millenia.

Cutting fresh meat **may** leave a material residue on the surface of the quartz tool which is unaffected by chemical cleaning. A larger sample size should be examined to confirm the consistency of this feature as a diagnostic trait. This was the most difficult activity to discern on quartz. There was generally little edge damage inflicted along the working edge during use and what little damage was present, was undiagnostic; it could easily be mistaken for trampling or some non-use related cause.

Unfortunately, no diagnostic use wear traces could be found on those quartz tools used for shell working or feather working. While hexagonal luminescent "plates" were noted away from the working edge, they were inconsistently present. It is also possible that when present, they may be removed by chemical cleaning. Care, therefore, should be taken to examine artefacts thoroughly prior to any acidic cleaning.

Those familiar with Keeley's method of microwear analysis on flint tools will note that there are certain similarities existing between microwear polishes on flint and quartz; a soft plant polish, for example, is extremely similar on both. The highly reflective surface of the quartz makes identification of the use wear polishes more difficult as the contrast between the dark flint and the corresponding bright polish is lacking. The "melted snow" quality seen on flint after working soaked antler is also present on quartz as are the "cracks" running through the polish after working bone. These cracks appear more frequently on the experimental quartz pieces than on those flint tools used for bone working. A "classic" wood polish (if indeed there is such a thing), is more diagnostic and easily recognizable on flint tools than on quartz.

A major difference between microwear polishes on quartz and flint is the presence of pits and linear grooves on almost all the experimental quartz tools. Their presence and abundance in relation to each other aid in identifying use. One of the problems in working with quartz tools is, that unless the tool has been used for more than five minutes, in most cases, it is extremely difficult to pinpoint use with 100% accuracy. Some attrition of the working edges and ridges is generally visible almost immediately after use begins, however, attrition looks the same no matter what material the quartz tool was used on. After such a short period of time, the use wear polish has not developed sufficiently to make it visible against the reflective background of the quartz. When microscopically examining the relatively dark surface of the flint, the beginning stages of a polish tend to distinguish themselves earlier on.

It should be remembered that the statistics presented in this study are used to strengthen the observational results obtained microscopically. These features, (except perhaps, for the comet-tailed pits and the cracks in the polish), are meaningless when examined in isolation. It is precisely the presence of certain features in association with other features and the optical qualities (which are not quantifiable here) which aid in determining the precise function of a quartz tool.

Microwear analysis on quartz is a relatively new endeavor which has yet to be fully explored. This monograph shows that microwear polish and other features are present on experimental quartz tools and that it is possible to identify the various activities for which experimental stone tools of quartz were used. The next step is to apply these findings to appropriate archaeological assemblages for testing. Microwear on quartz is not as easily recognizable as flint but, with further research and refinement, there is no reason to doubt that the technique will be valuable and informative on archaeological assemblages. Training of others interested in learning the technique requires patience and involves much time simply becoming "acquainted" with unused quartz surfaces microscopically. One drawback in utilizing microwear analysis on quartz is the inconsistency and unpredictability with which recognizable features are visible; so much of what is recognizable is determined by the surface texture of the quartz.

Even so, there is still much that can be done to refine and further explore polish formation and identification on quartz tools. It is a valuable tool and deserves further research effort. Future research would include focusing more attention on how these polishes form - sectioning tool edges with well developed polishes and employing x-ray diffraction and microprobe techniques to a variety of different polishes, both on stone and metal for comparison.

Emphasis should be turned towards an increased exploration of the effects of the various chemical cleaning techniques used and exactly what their effects are on the various polishes. While concern has been raised regarding the potential alteration of polishes due to chemical cleaning (see, for example, Unger-Hamilton 1983:245) and researchers are beginning to move towards less potentially destructive methods, there are no publications dealing specifically with the pros and cons of the various cleaning methods commonly used. The effects of soil acids on the surface of stone tools and ?subsequent alteration of the polishes (on flint as well as quartz) are benefitting from the beginnings of in depth investigations as well. Chapter two of this monograph discusses the recent developments in studying the post-depositional effects on microwear polish. It is possible that what we are finding on quartz tools now is a more realistic approximation of what we may expect to find in the archaeological assemblages once the effects of "time" are teased out. When all is said and done, use wear polishes may not be a diagnostic criteria and other features such as pitting, linear grooves and comet-tailed pits may survive to accurately tell the wear story on stone tools. Preliminary research using acids to microscopically etch the surface of quartz tools (Knutsson, pers. comm.) enhances the presence of these features. Further study and experiments hopefully will uncover which of these features can survive the ravages of time (soil movement and acids), and tell us, with a certain degree of accuracy, what these stone tools were used for.

Summary Table of Attribute Means

Material Worked	Pits	Striae	Linear Grooves	Comet-tailed Pits
All Grass Cutting	56.6	0.8	10.2	0
All Hide Scraping	39.0	3.4	4.7	0
All Wood Working	148.5	1.2	26.8	0
All Antler Working	93.6	3.3	21.0	38.9
All Bone Working	29.5	4.0	6.2	24.7

Table 23

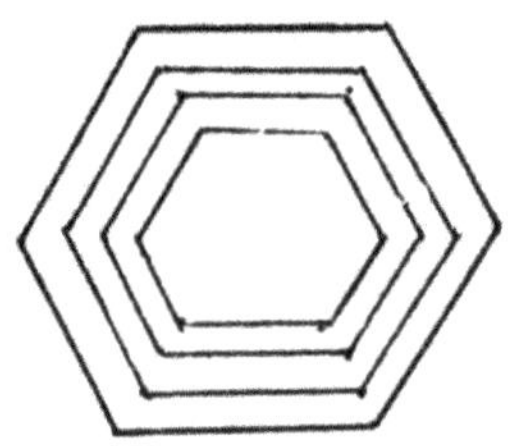

Figure 39

Appendix 1

Combined Attribute Table for Experimental Quartz Tools

Combined Attribute Table for Use Wear On Quartz

Material → / Attributes ↓	GRASS	SUN-DRIED & TANNED HIDE	FRESH HIDE	WOOD	DRY ANTLER	SOAKED ANTLER	BONE
BRIGHTNESS	BRIGHT	—	DULL	BRIGHT	—	BRIGHT	BRIGHT
LUSTER	METALLIC	—	—	METALLIC	—	METALLIC	GREASY
TEXTURE	SMOOTH	ROUGH	ROUGH	SMOOTH &/OR ROUGH	ROUGH	"MELTED SNOW"	SMOOTH & CRACKS
LINEAR GROOVES	YES	YES	FEW	YES	FEW	YES	YES
POLISH FORMATION	FAST	—	SLOW	VARIABLE	—	MODERATE	FAST
PITTING	YES	YES	FEW	YES	FEW	COMET-TAIL	COMET-TAIL
STRIATIONS	YES	YES	NO	YES	YES	YES	YES
EDGE DAMAGE	YES	YES	YES	YES	YES	YES	YES

Appendix 2

Attribute Tables for Experimental Quartz Tools Grouped by Material Worked

0 = Absent
1 = Present
2 = Smooth Surface Texture
3 = Rough Surface Texture

Scraping Sun-dried Hide

TIME ↓ / ATTRIBUTES →	SCARS	PITS	COMET-TAILED PITS	STRIAE	LINEAR GROOVES	RIDGE ABRASION	POLISH	EDGE ABRASION	EDGE ROUNDING	RIDGE ROUNDING	SURFACE TEXTURE	"MELTED SNOW"	"CRACKS" IN THE POLISH
One Minute	1	1	0	1	1	0	0	1	0	0	0	0	0
Five Minutes	1	1	0	1	1	1	0	1	0	0	1	0	0
Ten Minutes	1	1	0	1	1	1	0	1	1	0	1	0	0
Twenty Minutes	1	1	0	1	1	1	0	1	1	0	0	0	0
Sixty Minutes	1	1	0	1	1	1	0	1	1	0	0	0	0

1 = Presence
0 = Absence

Scraping Fresh Hide

TIME ↓ / ATTRIBUTES →	SCARS	PITS	COMET-TAILED PITS	STRIAE	LINEAR GROOVES	RIDGE ABRASION	POLISH	EDGE ABRASION	EDGE ROUNDING	RIDGE ROUNDING	SURFACE TEXTURE	"MELTED SNOW"	"CRACKS" IN THE POLISH
One Minute	1	0	0	0	0	0	0	0	0	0	0	0	0
Five Minutes	1	1	0	0	0	0	0	0	0	0	0	0	0
Ten Minutes	1	1	0	1	0	1	0	1	0	0	0	0	0
Twenty Minutes	1	1	0	1	1	0	0	1	0	0	0	0	0
Sixty Minutes													

1 = Presence
0 = Absence

Cutting Grass (Tilden Park) — ATTRIBUTES → / TIME ↓	SCARS	PITS	COMET-TAILED PITS	STRIAE	LINEAR GROOVES	RIDGE ABRASION	POLISH	EDGE ABRASION	EDGE ROUNDING	RIDGE ROUNDING	SURFACE TEXTURE	"MELTED SNOW"	"CRACKS" IN THE POLISH
One Minute	1	0	0	0	0	0	1	0	0	0	0	0	0
Five Minutes	1	1	0	0	1	0	1	0	1	0	0	0	0
Ten Minutes	1	1	0	1	1	1	1	0	1	0	0	0	0
Twenty Minutes	1	1	0	0	1	0	1	0	1	0	2	0	0
Sixty Minutes	1	1	0	1	1	1	1	0	1	1	3	0	0

1 = Presence
0 = Absence
2 = Smooth
3 = Rough

Cutting Grass (Drake's Bay) ATTRIBUTES → / TIME ↓	SCARS	PITS	COMET-TAILED PITS	STRIAE	LINEAR GROOVES	RIDGE ABRASION	POLISH	EDGE ABRASION	EDGE ROUNDING	RIDGE ROUNDING	SURFACE TEXTURE	"MELTED SNOW"	"CRACKS" IN THE POLISH
One Minute	1	0	0	0	1	1	0	0	0	0	0	0	0
Five Minutes	1	1	0	1	1	0	0	0	0	0	0	0	0
Ten Minutes	1	1	0	1	1	0	1	0	1	1	0	0	0
Twenty Minutes	1	1	0	1	1	0	1	1	1	0	2	0	0
Sixty Minutes	1	1	0	1	1	1	1	1	1	1	3	0	0

1 = Presence
0 = Absence
2 = Smooth
3 = Rough

Sawing Wood (Acacia)

TIME ↓ / ATTRIBUTES →	SCARS	PITS	COMET-TAILED PITS	STRIAE	LINEAR GROOVES	RIDGE ABRASION	POLISH	EDGE ABRASION	EDGE ROUNDING	RIDGE ROUNDING	SURFACE TEXTURE	"MELTED SNOW"	"CRACKS" IN THE POLISH
One Minute	1	1	0	1	1	1	0	1	0	0	0	0	0
Five Minutes	1	1	0	1	1	1	0	1	0	1	0	0	0
Ten Minutes	1	1	0	1	1	1	0	1	1	0	1	0	0
Twenty Minutes	1	1	0	1	1	1	1	1	1	0	0	0	0
Sixty Minutes	1	1	0	1	1	1	0	1	1	0	1	0	0

1 = Presence
0 = Absence

Sawing Wood (Eucalyptus)

TIME ↓ / ATTRIBUTES →	SCARS	PITS	COMET-TAILED PITS	STRIAE	LINEAR GROOVES	RIDGE ABRASION	POLISH	EDGE ABRASION	EDGE ROUNDING	RIDGE ROUNDING	SURFACE TEXTURE	"MELTED SNOW"	"CRACKS" IN THE POLISH
One Minute	1	1	0	1	1	1	0	1	0	0	0	0	0
Five Minutes	1	1	0	1	1	1	0	1	0	0	0	0	0
Ten Minutes	1	1	0	0	1	1	0	1	1	1	3	0	0
Twenty Minutes	1	1	0	0	1	1	1	1	1	1	0	0	0
Sixty Minutes	1	1	0	1	1	1	1	1	1	1	1	0	0

1 = Presence
0 = Absence
3 = Smooth

Sawing Wood (Oak)

TIME ↓ / ATTRIBUTES →	SCARS	PITS	COMET-TAILED PITS	STRIAE	LINEAR GROOVES	RIDGE ABRASION	POLISH	EDGE ABRASION	EDGE ROUNDING	RIDGE ROUNDING	SURFACE TEXTURE	"MELTED SNOW"	"CRACKS" IN THE POLISH
One Minute	1	1	0	1	0	0	0	1	0	0	0	0	0
Five Minutes	1	1	0	0	1	1	0	1	0	0	1	0	0
Ten Minutes	1	1	0	0	1	0	0	0	0	0	1	0	0
Twenty Minutes	1	1	0	0	1	0	0	0	1	1	0	0	0
Sixty Minutes	1	1	0	1	1	1	0	1	1	1	2	0	0

1 = Presence
0 = Absence
2 = Smooth

Sawing Wood (Pine)

TIME ↓ / ATTRIBUTES →	SCARS	PITS	COMET-TAILED PITS	STRIAE	LINEAR GROOVES	RIDGE ABRASION	POLISH	EDGE ABRASION	EDGE ROUNDING	RIDGE ROUNDING	SURFACE TEXTURE	"MELTED SNOW"	"CRACKS" IN THE POLISH
One Minute	0	1	0	1	1	1	0	1	0	0	0	0	0
Five Minutes													
Ten Minutes	1	1	0	0	1	1	0	1	1	0	1	0	0
Twenty Minutes	1	1	0	1	1	1	1	0	1	0	2	0	0
Sixty Minutes	1	1	0	1	1	1	1	0	1	0	0	0	0

1 = Presence
0 = Absence
2 = Smooth

Sawing Wood (Yew) — ATTRIBUTES → / TIME ↓	SCARS	PITS	COMET-TAILED PITS	STRIAE	LINEAR GROOVES	RIDGE ABRASION	POLISH	EDGE ABRASION	EDGE ROUNDING	RIDGE ROUNDING	SURFACE TEXTURE	"MELTED SNOW"	"CRACKS" IN THE POLISH
One Minute	1	1	0	1	1	1	0	1	0	0	0	0	0
Five Minutes	1	1	0	1	1	1	0	1	1	0	1	0	0
Ten Minutes	1	1	0	0	0	0	0	0	0	0	0	0	0
Twenty Minutes													
Sixty Minutes													

1 = Presence
0 = Absence

Scraping Wood (Acacia)

TIME / ATTRIBUTES	SCARS	PITS	COMET-TAILED PITS	STRIAE	LINEAR GROOVES	RIDGE ABRASION	POLISH	EDGE ABRASION	EDGE ROUNDING	RIDGE ROUNDING	SURFACE TEXTURE	"MELTED SNOW"	"CRACKS" IN THE POLISH
One Minute	1	1	0	0	0	0	0	1	0	0	0	0	0
Five Minutes	1	0	0	1	1	1	0	1	0	0	0	0	0
Ten Minutes	1	1	0	0	1	0	0	0	0	0	0	0	0
Twenty Minutes	1	1	0	1	1	0	0	1	0	0	0	0	0
Sixty Minutes	1	1	0	0	1	0	0	1	0	0	0	0	0

1 = Presence
0 = Absence

Scraping Wood (Eucalyptus)

TIME ↓ / ATTRIBUTES →	SCARS	PITS	COMET-TAILED PITS	STRIAE	LINEAR GROOVES	RIDGE ABRASION	POLISH	EDGE ABRASION	EDGE ROUNDING	RIDGE ROUNDING	SURFACE TEXTURE	"MELTED SNOW"	"CRACKS" IN THE POLISH
One Minute	1	0	0	0	1	0	0	0	0	0	0	0	0
Five Minutes	1	0	0	1	1	0	0	1	0	0	0	0	0
Ten Minutes	1	1	0	1	1	0	0	1	0	0	1	0	0
Twenty Minutes	1	1	0	0	1	0	0	1	1	0	0	0	0
Sixty Minutes	1	1	0	1	1	1	1	0	1	1	0	0	0

1 = Presence
0 = Absence

Scraping Wood (Oak)

TIME ↓ / ATTRIBUTES →	SCARS	PITS	COMET-TAILED PITS	STRIAE	LINEAR GROOVES	RIDGE ABRASION	POLISH	EDGE ABRASION	EDGE ROUNDING	RIDGE ROUNDING	SURFACE TEXTURE	"MELTED SNOW"	"CRACKS" IN THE POLISH
One Minute	1	1	0	0	1	0	0	1	0	0	0	0	0
Five Minutes	1	1	0	1	1	0	0	1	0	0	0	0	0
Ten Minutes	1	1	0	1	1	0	0	0	0	0	0	0	0
Twenty Minutes	1	1	0	1	1	1	1	1	1	0	2	0	0
Sixty Minutes	1	1	0	1	1	0	1	1	1	0	0	0	0

1 = Presence
0 = Absence
2 = Smooth

Sawing Soaked Antler

TIME ↓ / ATTRIBUTES →	SCARS	PITS	COMET-TAILED PITS	STRIAE	LINEAR GROOVES	RIDGE ABRASION	POLISH	EDGE ABRASION	EDGE ROUNDING	RIDGE ROUNDING	SURFACE TEXTURE	"MELTED SNOW"	"CRACKS" IN THE POLISH
One Minute	1	1	0	1	1	1	0	1	0	0	0	0	0
Five Minutes	1	1	1	1	1	1	1	1	0	0	1	1	0
Ten Minutes	1	1	1	1	1	1	1	1	1	1	2	1	0
Twenty Minutes	1	1	1	1	1	1	1	1	1	1	0	1	0
Sixty Minutes	1	1	1	1	1	1	1	1	1	1	0	1	0

1 = Presence
0 = Absence
2 = Smooth

Sawing Dry Antler

TIME ↓ / ATTRIBUTES →	SCARS	PITS	COMET-TAILED PITS	STRIAE	LINEAR GROOVES	RIDGE ABRASION	POLISH	EDGE ABRASION	EDGE ROUNDING	RIDGE ROUNDING	SURFACE TEXTURE	"MELTED SNOW"	"CRACKS" IN THE POLISH
One Minute	1	1	0	1	1	1	0	1	0	0	0	0	0
Five Minutes	1	1	1	1	1	0	1	0	0	1	0	0	0
Ten Minutes	1	1	0	1	1	1	0	1	0	0	0	0	0
Twenty Minutes	1	1	0	1	1	1	0	1	0	0	0	0	0
Sixty Minutes	1	1	0	1	1	1	0	1	0	0	0	0	0

1 = Presence
0 = Absence

Scraping Soaked Antler

TIME → / ATTRIBUTES ↑	SCARS	PITS	COMET-TAILED PITS	STRIAE	LINEAR GROOVES	RIDGE ABRASION	POLISH	EDGE ABRASION	EDGE ROUNDING	RIDGE ROUNDING	SURFACE TEXTURE	"MELTED SNOW"	"CRACKS" IN THE POLISH
One Minute	1	0	0	1	1	0	0	1	0	0	0	0	0
Five Minutes	1	1	1	1	1	1	1	1	0	0	0	0	0
Ten Minutes	1	1	0	1	1	1	0	1	0	0	1	0	0
Twenty Minutes	1	1	0	1	1	1	0	1	0	0	0	1	0
Sixty Minutes	1	1	1	1	1	1	1	1	1	0	2	1	0

1 = Presence
0 = Absence
2 = Smooth

Graving Soaked Antler

TIME ↓ / ATTRIBUTES →	SCARS	PITS	COMET-TAILED PITS	STRIAE	LINEAR GROOVES	RIDGE ABRASION	POLISH	EDGE ABRASION	EDGE ROUNDING	RIDGE ROUNDING	SURFACE TEXTURE	"MELTED SNOW"	"CRACKS" IN THE POLISH
One Minute	1	0	1	1	1	1	0	1	0	0	0	0	0
Five Minutes	1	1	1	1	1	0	1	1	1	0	1	1	0
Ten Minutes	1	0	1	1	1	1	1	1	1	0	0	0	0
Twenty Minutes	1	1	0	1	1	1	1	1	1	0	0	1	0
Sixty Minutes	1	1	1	1	1	1	1	0	1	1	2	1	0

1 = Presence
0 = Absence
2 = Smooth

Sawing Bone — ATTRIBUTES → / TIME ↓	SCARS	PITS	COMET-TAILED PITS	STRIAE	LINEAR GROOVES	RIDGE ABRASION	POLISH	EDGE ABRASION	EDGE ROUNDING	RIDGE ROUNDING	SURFACE TEXTURE	"MELTED SNOW"	"CRACKS" IN THE POLISH
One Minute	1	0	1	1	1	1	0	0	0	0	0	0	0
Five Minutes	1	1	1	1	1	1	1	1	0	1	0	0	0
Ten Minutes	1	1	1	1	1	0	1	1	1	1	2	0	0
Twenty Minutes	1	1	1	1	1	1	1	1	1	1	1	0	0
Sixty Minutes	1	1	1	1	1	1	0	1	1	1	3	0	0

1 = Presence
0 = Absence
2 = Smooth
3 = Rough

Scraping Bone

TIME ↓ / ATTRIBUTES →	SCARS	PITS	COMET-TAILED PITS	STRIAE	LINEAR GROOVES	RIDGE ABRASION	POLISH	EDGE ABRASION	EDGE ROUNDING	RIDGE ROUNDING	SURFACE TEXTURE	"MELTED SNOW"	"CRACKS" IN THE POLISH
One Minute	1	1	1	0	0	0	0	1	1	0	0	0	0
Five Minutes	1	1	1	1	0	0	1	1	1	0	0	0	1
Ten Minutes	1	1	1	1	1	1	1	1	1	0	0	0	1
Twenty Minutes	1	1	1	1	1	0	1	1	1	0	2	0	1
Sixty Minutes	1	1	1	1	1	1	0	1	1	1	0	0	1

1 = Presence
0 = Absence
2 = Smooth

Graving Fresh Bone

TIME / ATTRIBUTES	SCARS	PITS	COMET-TAILED PITS	STRIAE	LINEAR GROOVES	RIDGE ABRASION	POLISH	EDGE ABRASION	EDGE ROUNDING	RIDGE ROUNDING	SURFACE TEXTURE	"MELTED SNOW"	"CRACKS" IN THE POLISH
One Minute	1	1	1	0	0	0	0	1	1	0	1	0	0
Five Minutes	1	0	1	0	0	1	1	1	1	0	1	0	0
Ten Minutes	1	1	1	1	1	1	1	0	1	0	0	0	1
Twenty Minutes	1	0	1	1	0	1	0	0	1	1	2	0	0
Sixty Minutes	1	1	1	1	1	1	1	0	1	1	0	0	1

1 = Presence
0 = Absence
2 = Smooth

Appendix 3

Attribute Counts and Statistics for Experimental Quartz Tools

MINIMUM, MAXIMUM, MEAN AND STANDARD DEVIATIONS FOR ATTRIBUTE COUNTS

CUTTING ALL GRASS ** TIMES COMBINED

N=105

	C.T. PITS	STRIAE	PITS	LINEAR GROOVES
min	0	0	0	0
max	0	10	500	75
mean	0	.84	56.55	10.19
s.d.	0	1.67	84.63	12.57

CUTTING GRASS ** DRAKE'S BAY ** TIMES COMBINED

N=50

	C.T. PITS	STRIAE	PITS	LINEAR GROOVES
min	0	0	0	0
max	0	8	500	75
mean	0	1.14	89.48	14.86
s.d.	0	1.67	108.45	14.87

CUTTING GRASS ** TILDEN PARK ** TIMES COMBINED

N=55

	C.T. PITS	STRIAE	PITS	LINEAR GROOVES
min	0	0	0	0
max	0	10	175	39
mean	0	.56	26.62	5.95
s.d.	0	1.62	33.16	7.93

CUTTING ALL GRASS ** ONE MINUTE

N=20

C.T. PITS	STRIAE	PITS	LINEAR GROOVES
NONE	min 0	min 0	min 0
	max 1	max 22	max 8
	mean .1	mean 5.9	mean 1.9
	s.d. .3	s.d. 6.35	s.d. 2.72

CUTTING ALL GRASS ** FIVE MINUTES

N=20

C.T. PITS	STRIAE	PITS	LINEAR GROOVES
NONE	min 0	min 0	min 0
	max 10	max 60	max 21
	mean .8	mean 22.95	mean 7.25
	s.d. 2.20	s.d. 18.43	s.d. 6.20

CUTTING ALL GRASS ** TEN MINUTES

N=20

C.T. PITS	STRIAE	PITS	LINEAR GROOVES
NONE	min 0	min 2	min 0
	max 6	max 130	max 42
	mean .7	mean 36.9	mean 8.15
	s.d. 1.45	s.d. 37.54	s.d. 9.79

CUTTING ALL GRASS ** TWENTY MINUTES

N=25

C.T. PITS	STRIAE	PITS	LINEAR GROOVES
NONE	min 0	min 0	min 0
	max 4	max 300	max 46
	mean .72	mean 85.08	mean 12.56
	s.d. 1.04	s.d. 85.59	s.d. 12.34

CUTTING ALL GRASS ** SIXTY MINUTES

N=20

C.T. PITS	STRIAE	PITS	LINEAR GROOVES
NONE	min 0	min 12	min 0
	max 8	max 500	max 75
	mean 1.9	mean 124.8	mean 20.5
	s.d. 2.12	s.d. 129.99	s.d. 17.39

CUTTING GRASS ** DRAKE'S BAY ** ONE MINUTE

N=10

C.T. PITS	STRIAE	PITS	LINEAR GROOVES
NONE	NONE	min 0	min 0
		max 13	max 8
		mean 5	mean 1.5
		s.d. 5.16	mean 2.38

CUTTING GRASS ** DRAKE'S BAY ** 5 MINUTES

N=10

C.T. PITS	STRIAE	PITS	LINEAR GROOVES
NONE	min 0	min 14	min 2
	max 2	max 60	max 21
	mean .6	mean 34.9	mean 11.1
	s.d. .8	s.d. 14.58	s.d. 6.53

CUTTING GRASS ** DRAKE'S BAY ** 10 MINUTES

N=10

C.T. PITS	STRIAE	PITS	LINEAR GROOVES
NONE	min 0	min 2	min 1
	max 6	max 130	max 42
	mean 1.3	mean 51.3	mean 10.9
	s.d. 1.85	s.d. 45.06	s.d. 12.13

CUTTING GRASS ** DRAKE'S BAY ** 20 MINUTES

N=10

C.T. PITS	STRIAE	PITS	LINEAR GROOVES
NONE	min 0	min 80	min 6
	max 4	max 300	max 46
	mean 1.7	mean 170.2	mean 24.1
	s.d. 1.00	s.d. 71.08	s.d. 11.37

CUTTING GRASS ** DRAKE'S BAY ** 60 MINUTES

N=10

C.T. PITS	STRIAE	PITS	LINEAR GROOVES
NONE	min 0	min 50	min 11
	max 8	max 500	max 75
	mean 2.1	mean 186	mean 26.7
	s.d. 2.47	s.d. 155.18	s.d. 18.61

CUTTING GRASS ** TILDEN PARK ** ONE MINUTE

N=10

C.T. PITS	STRIAE	PITS	LINEAR GROOVES
NONE	min 0	min 0	min 0
	max 1	max 22	max 8
	mean .2	mean 6.8	mean 2.3
	s.d. .4	s.d. 7.24	s.d. 2.97

CUTTING GRASS ** TILDEN PARK ** 5 MINUTES

N=10

C.T. PITS	STRIAE	PITS	LINEAR GROOVES
NONE	min 0	min 0	min 0
	max 10	max 45	max 7
	mean 1	mean 11	mean 3.4
	s.d. 3	s.d. 13.45	s.d. 2.15

CUTTING GRASS ** TILDEN PARK ** 10 MINUTES

N=10

C.T. PITS	STRIAE	PITS	LINEAR GROOVES
NONE	min 0	min 2	min 0
	max 1	max 67	max 17
	mean .1	mean 22.5	mean 5.4
	s.d. .3	s.d. 19.34	s.d. 5.43

CUTTING GRASS ** TILDEN PARK ** 20 MINUTES

N=15

C.T. PITS	STRIAE	PITS	LINEAR GROOVES
NONE	min 0	min 0	min 0
	max 1	max 100	max 13
	mean .067	mean 28.33	mean 4.87
	s.d. .25	s.d. 28.14	s.d. 4.44

CUTTING GRASS ** TILDEN PARK ** 60 MINUTES

N=10

C.T. PITS	STRIAE	PITS	LINEAR GROOVES
NONE	min 0	min 12	min 0
	max 5	max 175	max 39
	mean 1.7	mean 63.6	mean 14.3
	s.d. 1.68	s.d. 47.19	s.d. 13.48

SCRAPING ALL HIDES ** ALL TIMES COMBINED

N=125

C.T. PITS	STRIAE	PITS	LINEAR GROOVES
NONE	min 0	min 0	min 0
	max 60	max 999	max 38
	mean 3.34	mean 65.49	mean 4.48
	s.d. 8.60	s.d. 148.29	s.d. 7.13

SCRAPING ALL HIDES...PIECE WITH VOLCANIC ASH DELETED
ALL TIMES COMBINED

N=120

C.T. PITS	STRIAE	PITS	LINEAR GROOVES
NONE	min 0	min 0	min 0
	max 60	max 300	max 38
	mean 3.39	mean 39.07	mean 4.66
	s.d. 8.16	s.d. 54.86	s.d. 7.23

ALL HIDE SCRAPING ** ONE MINUTE

N=25

C.T. PITS	STRIAE	PITS	LINEAR GROOVES
NONE	min 0	min 0	min 0
	max 5	max 300	max 38
	mean .68	mean 34.6	mean 6.52
	s.d. 1.49	s.d 58.69	s.d. 10.09

ALL HIDE SCRAPING ** 5 MINUTES

N=25

C.T. PITS	STRIAE	PITS	LINEAR GROOVES
NONE	min 0	min 0	min 0
	max 11	max 250	max 22
	mean .88	mean 40.32	mean 3.44
	s.d. 2.50	s.d. 54.15	s.d 5.49

ALL HIDE SCRAPING ** 10 MINUTES

N=30

C.T. PITS	STRIAE	PITS	LINEAR GROOVES
NONE	min 0	min 0	min 0
	max 9	max 999	max 8
	mean 1.6	mean 135.6	mean 1.6
	s.d. 2.54	s.d. 271.45	s.d. 2.30

ALL HIDE SCRAPING ** 20 MINUTES

N=25

C.T. PITS	STRIAE	PITS	LINEAR GROOVES
NONE	min 0	min 0	min 0
	max 30	max 71	max 13
	mean 4.68	mean 24.2	mean 2.88
	s.d. 9.20	s.d. 22.16	s.d. 4.19

ALL HIDE SCRAPING ** 60 MINUTES

N=20

C.T. PITS	STRIAE	PITS	LINEAR GROOVES
NONE	min 0	min 7	min 0
	max 60	max 280	max 30
	mean 10.7	mean 82.05	mean 9.55
	s.d. 14.03	s.d. 81.02	s.d. 8.69

FRESH HIDE SCRAPING ** TIMES COMBINED

N=25

C.T. PITS	STRIAE	PITS	LINEAR GROOVES
NONE	min 0	min 0	min 0
	max 3	max 999	max 6
	mean .44	mean 159.76	mean .84
	s.d. .94	s.d. 291.44	s.d. 1.59

FRESH HIDE SCRAPING WITHOUT VOLCANIC ASH PIECE *** TIMES COMBINED

N=20

C.T. PITS	STRIAE	PITS	LINEAR GROOVES
NONE	NONE	min 0	min 0
		max 65	max 6
		mean 24.75	mean 1
		s.d. 20.63	s.d. 1.73

SCRAPING SUN-DRIED HIDE ** TIMES COMBINED

N=75

C.T. PITS	STRIAE	PITS	LINEAR GROOVES
NONE	min 0	min 0	min 0
	max 12	max 300	max 38
	mean 1.53	mean 51.03	mean 6.48
	s.d. 2.95	s.d. 65.14	s.d. 8.43

SCRAPING TANNED HIDE ** TIMES COMBINED

N=25

C.T. PITS	STRIAE	PITS	LINEAR GROOVES
NONE	min 0	min 0	min 0
	max 60	max 46	max 12
	mean 11.68	mean 14.64	mean 2.12
	s.d. 14.34	s.d. 13.54	s.d. 2.85

SCRAPING FRESH HIDE ** ONE MINUTE

N=5

C.T. PITS	STRIAE	PITS	LINEAR GROOVES
NONE	NONE	NONE	NONE

SCRAPING FRESH HIDE ** 5 MINUTES

N=5

C.T. PITS	STRIAE	PITS	LINEAR GROOVES
NONE	NONE	min 26	min 0
		max 60	max 6
		mean 41.4	mean 3.2
		s.d. 11.02	s.d. 2.14

SCRAPING FRESH HIDE ** 10 MINUTES

N=10

C.T. PITS	STRIAE	PITS	LINEAR GROOVES
NONE	min 0	min 8	min 0
	max 3	max 999	max 1
	mean 1.1	mean 359.1	mean .2
	s.d. 1.22	s.d. 381.24	s.d. .4

SCRAPING FRESH HIDE ** 10 MINUTES ** WITHOUT ASH

N=5

C.T. PITS	STRIAE	PITS	LINEAR GROOVES
NONE	NONE	min 8	min 0
		max 37	max 1
		mean 18.4	mean .2
		s.d. 10.05	s.d. .4

SCRAPING FRESH HIDE ** Q#12 VOLCANIC ASH ADDED

N=1

C.T. PITS	STRIAE	PITS	LINEAR GROOVES
NONE	min 1	min 300	min 0
	max 3	max 900	max 1
	mean 2.2	mean 699.8	mean .2
	s.d. .75	s.d. 241.74	s.d. .4

SCRAPING FRESH HIDE ** 20 MINUTES

N=5

C.T. PITS	STRIAE	PITS	LINEAR GROOVES
NONE	NONE	min 14	min 0
		max 65	max 2
		mean 39.2	mean .6
		s.d. 18.45	s.d. .8

SCRAPING SUN-DRIED HIDE ** ONE MINUTE

N=15

C.T. PITS	STRIAE	PITS	LINEAR GROOVES
NONE	min 0	min 0	min 0
	max 5	max 300	max 38
	mean .67	mean 49.67	mean 9.8
	s.d. 1.45	s.d. 70.69	s.d. 11.58

SCRAPING SUN-DRIED HIDE ** 5 MINUTES

N=15

C.T. PITS	STRIAE	PITS	LINEAR GROOVES
NONE	min 0	min 6	min 0
	max 11	max 250	max 22
	mean 1.47	mean 53.4	mean 4.67
	s.d. 3.09	s.d. 64.29	s.d. 6.58

SCRAPING SUN-DRIED HIDE **10 MINUTES

N=15

C.T. PITS	STRIAE	PITS	LINEAR GROOVES
NONE	min 0	min 0	min 0
	max 4	max 80	max 8
	mean .33	mean 24.73	mean 2.27
	s.d. 1.01	s.d. 22.21	s.d. 4.78

SCRAPING SUN-DRIED HIDE ** 20 MINUTES

N=15

C.T. PITS	STRIAE	PITS	LINEAR GROOVES
NONE	min 0	min 0	min 0
	max 6	max 71	max 13
	mean 1	mean 23.93	mean 3.87
	s.d. 2	s.d. 22.46	s.d. 4.79

SCRAPING SUN-DRIED HIDE ** 60 MINUTES

N=15

C.T. PITS	STRIAE	PITS	LINEAR GROOVES
NONE	min 0	min 15	min 0
	max 12	max 280	max 30
	mean 4.2	mean 103.4	mean 11.8
	s.d. 4.13	s.d. 83.14	s.d. 8.94

SCRAPING TANNED HIDE ** ONE MINUTE

N=5

C.T. PITS	STRIAE	PITS	LINEAR GROOVES
NONE	min 0	min 0	min 0
	max 5	max 46	max 12
	mean 1.4	mean 24	mean 3.2
	s.d. 1.96	s.d. 15.40	s.d. 4.53

SCRAPING TANNED HIDE ** 5 MINUTES

N=5

C.T. PITS	STRIAE	PITS	LINEAR GROOVES
NONE	NONE	NONE	NONE

SCRAPING TANNED HIDE ** 10 MINUTES

N=5

C.T. PITS	STRIAE	PITS	LINEAR GROOVES
NONE	min 4	min 8	min 0
	max 9	max 30	max 4
	mean 6.4	mean 21.2	mean 2.4
	s.d. 2.06	s.d. 7.47	s.d. 1.62

SCRAPING TANNED HIDE ** 20 MINUTES

N=5

C.T. PITS	STRIAE	PITS	LINEAR GROOVES
NONE	min 2	min 0	min 0
	max 30	max 35	max 8
	mean 20.4	mean 10	mean 2.2
	s.d. 10.09	s.d. 13.25	s.d. 3.12

SCRAPING TANNED HIDE ** 60 MINUTES

N=5

C.T. PITS	STRIAE	PITS	LINEAR GROOVES
NONE	min 19	min 7	min 1
	max 60	max 33	max 4
	mean 30.2	mean 18	mean 2.8
	s.d. 15.12	s.d. 8.41	s.d. 1.17

ALL ANTLER WORKING (SANS DRY) ** ALL TIMES COMBINED

N=180

C.T. PITS	STRIAE	PITS	LINEAR GROOVES
min 0	min 0	min 0	min 0
max 999	max 30	max 999	max 400
mean 38.99	mean 3.33	mean 93.56	mean 20.99
s.d. 93.99	s.d. 4.97	s.d. 170.34	s.d. 44.01

ALL ANTLER WORKING ** ALL TIMES COMBINED

N=105

C.T. PITS	STRIAE	PITS	LINEAR GROOVES
min 0	min 0	min 0	min 0
max 999	max 30	max 999	max 400
mean 34.24	mean 3.29	max 85.94	max 19.32
s.d. 88.45	s.d. 4.72	s.d. 161.07	s.d. 41.53

ALL ANTLER WORK ** ONE MINUTE

N=25

C.T. PITS	STRIAE	PITS	LINEAR GROOVES
min 0	min 0	min 0	min 0
max 4	max 11	max 100	max 60
mean .16	mean 1.48	mean 23.12	mean 7.48
s.d. .78	s.d. 2.74	s.d. 104.42	s.d. 12.54

ALL ANTLER WORK ** 5 MINUTES

N=25

C.T. PITS	STRIAE	PITS	LINEAR GROOVES
min 0	min 0	min 0	min 0
max 100	max 30	max 500	max 150
mean 17.56	mean 4.72	mean 79	mean 24.16
s.d. 29.98	s.d. 7.23	s.d. 104.42	s.d. 34.44

ALL ANTLER WORK ** 10 MINUTES

N=20

C.T. PITS	STRIAE	PITS	LINEAR GROOVES
min 0	min 0	min 25	min 6
max 250	max 15	max 999	max 400
mean 47.55	mean 4.9	mean 248.1	mean 69.2
s.d. 68.67	s.d. 3.63	s.d. 242.18	s.d. 103.00

ALL ANTLER WORK ** 20 MINUTES

N=15

C.T. PITS	STRIAE	PITS	LINEAR GROOVES
min 0	min 1	min 25	min 0
max 130	max 18	max 350	max 8
mean 18.6	mean 5.27	mean 122.47	mean 4.67
s.d. 35.54	s.d. 4.23	s.d. 89.84	s.d. 2.77

ALL ANTLER WORK ** 60 MINUTES ** SANS DRY

N=25

C.T. PITS	STRIAE	PITS	LINEAR GROOVES
min 7	min 0	min 4	min 4
max 170	max 25	max 350	max 55
mean 67.12	mean 3.8	mean 85.36	mean 15.92
s.d. 40.10	s.d. 4.8	s.d. 78.29	s.d. 11.41

SOAKED ANTLER GRAVING ** TIMES COMBINED

N=65

C.T. PITS	STRIAE	PITS	LINEAR GROOVES
min 0	min 0	min 0	min 0
max 300	max 20	max 200	max 120
mean 34.6	mean 1.91	mean 30.49	mean 8.35
s.d. 62.28	s.d. 3.21	s.d. 39.62	s.d. 17.13

SOAKED ANTLER GRAVING ** ONE MINUTE

N=15

C.T. PITS	STRIAE	PITS	LINEAR GROOVES
min 0	min 0	min 0	min 0
max 14	max 2	max 60	max 25
mean 1.8	mean .33	mean 11.27	mean 4.13
s.d. 3.85	s.d. .59	s.d. 17.60	s.d. 7.27

SOAKED ANTLER GRAVING ** 5 MINUTES

N=15

C.T. PITS	STRIAE	PITS	LINEAR GROOVES
min 0	min 0	min 0	min 0
max 45	max 12	max 200	max 60
mean 12.2	mean 2.2	mean 41.53	mean 12.47
s.d. 13.13	s.d. 3.04	s.d. 58.90	s.d. 16.56

SOAKED ANTLER GRAVING ** 10 MINUTES

N=15

C.T. PITS	STRIAE	PITS	LINEAR GROOVES
min 0	min 0	min 6	min 0
max 300	max 20	max 150	max 13
mean 44.6	mean 2.47	mean 31.87	mean 4
s.d. 78.32	s.d. 4.98	s.d. 35.96	s.d. 4.05

SOAKED ANTLER GRAVING ** 20 MINUTES

N=15

C.T. PITS	STRIAE	PITS	LINEAR GROOVES
min 2	min 0	min 3	min 1
max 300	max 10	max 100	max 120
mean 56	mean 2.47	mean 32.47	mean 11.8
s.d. 78.65	s.d. 2.58	s.d. 32.17	s.d. 29.04

SOAKED ANTLER GRAVING ** 60 MINUTES

N=5

C.T. PITS	STRIAE	PITS	LINEAR GROOVES
min 65	min 1	min 25	min 5
max 170	max 4	max 80	max 24
mean 106	mean 2.4	mean 45	mean 11.4
s.d. 37.34	s.d. 1.02	s.d. 19.49	s.d. 6.80

SOAKED ANTLER SAWING ** ALL TIMES COMBINED

N=180

C.T. PITS	STRIAE	PITS	LINEAR GROOVES
min 0	min 0	min 6	min 0
max 999	max 25	max 999	max 400
mean 58.31	mean 4.1	mean 125.69	mean 36.74
s.d. 132.61	s.d. 4.57	s.d. 168.38	s.d. 64.12

SAWING SOAKED ANTLER ** 60 MINUTES

N=10

C.T. PITS	STRIAE	PITS	LINEAR GROOVES
min 7	min 0	min 19	min 4
max 150	max 25	max 200	max 55
mean 58.6	mean 4.7	mean 113.6	mean 18.3
s.d. 38.96	s.d 7.11	s.d. 61.34	s.d. 14.81

SAWING SOAKED ANTLER ** 96 MINUTES

N=5

C.T. PITS	STRIAE	PITS	LINEAR GROOVES
min 200	min 3	min 40	min 1
max 999	max 10	max 300	max 70
mean 389.8	mean 5.8	mean 103	mean 23.8
s.d. 306.85	s.d. 2.4	s.d. 99.48	s.d. 24.84

SCRAPING SOAKED ANTLER ** TIMES COMBINED

N=45

C.T. PITS	STRIAE	PITS	LINEAR GROOVES
min 0	min 0	min 0	min 0
max 100	max 30	max 999	max 80
mean 15.29 9	mean 4.2	mean 134.67	mean 14.73
s.d. 27.70	s.d. 6.84	s.d. 246.26	s.d. 16.54

SCRAPING SOAKED ANTLER ** ONE MINUTE

N=10

C.T. PITS	STRIAE	PITS	LINEAR GROOVES
min 0	min 0	min 0	min 0
max 4	max 3	max 60	max 25
mean .4	mean .8	mean 22.8	mean 7.2
s.d. 1.2	s.d. 1.25	s.d. 19.22	s.d. 8.05

SAWING SOAKED ANTLER ** 60 MINUTES

N=10

C.T. PITS	STRIAE	PITS	LINEAR GROOVES
min 7	min 0	min 19	min 4
max 150	max 25	max 200	max 55
mean 58.6	mean 4.7	mean 113.6	mean 18.3
s.d. 38.96	s.d 7.11	s.d. 61.34	s.d. 14.81

SAWING SOAKED ANTLER ** 96 MINUTES

N=5

C.T. PITS	STRIAE	PITS	LINEAR GROOVES
min 200	min 3	min 40	min 1
max 999	max 10	max 300	max 70
mean 389.8	mean 5.8	mean 103	mean 23.8
s.d. 306.85	s.d. 2.4	s.d. 99.48	s.d. 24.84

SCRAPING SOAKED ANTLER ** TIMES COMBINED

N=45

C.T. PITS	STRIAE	PITS	LINEAR GROOVES
min 0	min 0	min 0	min 0
max 100	max 30	max 999	max 80
mean 15.29 9	mean 4.2	mean 134.67	mean 14.73
s.d. 27.70	s.d. 6.84	s.d. 246.26	s.d. 16.54

SCRAPING SOAKED ANTLER ** ONE MINUTE

N=10

C.T. PITS	STRIAE	PITS	LINEAR GROOVES
min 0	min 0	min 0	min 0
max 4	max 3	max 60	max 25
mean .4	mean .8	mean 22.8	mean 7.2
s.d. 1.2	s.d. 1.25	s.d. 19.22	s.d. 8.05

SCRAPING SOAKED ANTLER ** 5 MINUTES

N=10

C.T. PITS	STRIAE	PITS	LINEAR GROOVES
min 0	min 0	min 0	min 0
max 55	max 30	max 80	max 25
mean 9.2	mean 7	mean 32.7	mean 6.5
s.d. 17.69	s.d. 10.43	s.d. 25.86	s.d. 6.95

SCRAPING SOAKED ANTLER ** 10 MINUTES

N=5

C.T. PITS	STRIAE	PITS	LINEAR GROOVES
min 0	min 0	min 120	min 6
max 30	max 7	max 350	max 40
mean 6	mean 4.6	mean 206	mean 22.8
s.d. 12	s.d. 2.06	s.d. 84.05	s.d. 10.87

SCRAPING SOAKED ANTLER ** 20 MINUTES

N=5

C.T. PITS	STRIAE	PITS	LINEAR GROOVES
NONE	min 1	NONE	min 5
	max 6		max 25
	mean 4		mean 12.8
	s.d. 1.67		s.d. 8.13

SCRAPING SOAKED ANTLER ** 60 MINUTES

N=10

C.T. PITS	STRIAE	PITS	LINEAR GROOVES
min 10	min 1	min 4	min 4
max 100	max 8	max 350	max 30
mean 56.2	mean 3.6	mean 77.3	mean 15.8
s.d. 29.29	s.d. 2.15	s.d. 98.48	s.d. 8.19

SCRAPING SOAKED ANTLER ** 96 MINUTES

N=5

C.T. PITS	STRIAE	PITS	LINEAR GROOVES
NONE	min 0	min 0	min 0
	max 30	max 999	max 80
	mean 6.4	mean 616.4	mean 38
	s.d. 11.81	s.d. 469.36	s.d. 32.49

SAWING DRY ANTLER ** ALL TIMES COMBINED

N=25

C.T. PITS	STRIAE	PITS	LINEAR GROOVES
NONE	min 0	min 0	min 0
	max 7	max 70	max 20
	mean 3	mean 31.12	mean 7.32
	s.d. 2.23	s.d. 20.07	s.d. 5.55

SAWING DRY ANTLER ** ONE MINUTE

N=5

C.T. PITS	STRIAE	PITS	LINEAR GROOVES
NONE	min 0	min 8	min 2
	max 4	max 50	max 20
	mean 2.8	mean 27.8	mean 11.4
	s.d. 1.47	s.d. 15.10	s.d. 7.09

SAWING DRY ANTLER ** 5 MINUTES

N=5

C.T. PITS	STRIAE	PITS	LINEAR GROOVES
NONE	min 0	min 0	min 0
	max 7	max 50	max 10
	mean 2	mean 14	mean 3
	s.d. 2.76	s.d. 19.59	s.d. 4

SAWING DRY ANTLER ** 10 MINUTES

N=5

C.T. PITS	STRIAE	PITS	LINEAR GROOVES
NONE	min 3	min 15	min 8
	max 7	max 70	max 18
	mean 4.4	mean 46.4	mean 12.6
	s.d. 1.49	s.d. 21.21	s.d. 3.32

SAWING DRY ANTLER ** 20 MINUTES

N=5

C.T. PITS	STRIAE	PITS	LINEAR GROOVES
NONE	min 0	min 20	min 4
	max 7	max 50	max 6
	mean 2.6	mean 41	mean 4.8
	s.d. 2.42 1	s.d. 11.14	s.d. .75

SAWING DRY ANTLER ** 60 MINUTES

N=5

C.T. PITS	STRIAE	PITS	LINEAR GROOVES
NONE	min 1	min 10	min 4
	max 6	max 50	max 6
	mean 3.2	mean 26.4	mean 4.8
	s.d. 1.94	s.d. 13.03	s.d. .75

ALL WOOD ** ALL TIMES COMBINED

N=200

C.T. PITS	STRIAE	PITS	LINEAR GROOVES
NONE	min 0	min 0	min 0
	max 15	max 999	max 250
	mean 1.21	mean 148.45	mean 26.79
	s.d. 2.09	s.d. 217.74	s.d. 33.74

ALL WOOD WORKING ** ONE MINUTE

N=40

C.T. PITS	STRIAE	PITS	LINEAR GROOVES
NONE	min 0	min 0	min o
	max 8	max 400	max 45
	mean .4	mean 48.7	mean 11.125
	s.d. 1.36	s.d. 85.03	s.d. 14.69

ALL WOOD WORKING ** 5 MINUTES

N=35

C.T. PITS	STRIAE	PITS	LINEAR GROOVES
NONE	min 0	min 5	min 0
	max 6	max 500	max 65
	mean 1.14	mean 89.46	mean 20.34
	s.d. 1.5	s.d. 104.08	s.d. 18.61

ALL WOOD WORKING ** 10 MINUTES

N=45

C.T. PITS	STRIAE	PITS	LINEAR GROOVES
NONE	min 0	min 0	min 0
	max 7	max 999	max 95
	mean 1.02	mean 132.33	mean 22.47
	s.d. 1.51	s.d. 187.51	s.d. 22.99

ALL WOOD WORKING ** 20 MINUTES

N=40

C.T. PITS	STRIAE	PITS	LINEAR GROOVES
NONE	min 0	min 8	min 0
	max 7	max 999	max 250
	mean 1.43	mean 169.78	mean 44.2
	s.d. 1.83	s.d. 208.32	s.d. 55.13

ALL WOOD WORKING ** 60 MINUTES

N=40

C.T. PITS	STRIAE	PITS	LINEAR GROOVES
NONE	min 0	min 30	min 3
	max 15	max 999	max 150
	mean 2.05	mean 296.6	mean 33.55
	s.d. 3.23	s.d. 318.91	s.d. 29.24

ALL WOOD SAWING ** ALL TIMES COMBINED

N=125

C.T. PITS	STRIAE	PITS	LINEAR GROOVES
NONE	min 0	min 0	min 0
	max 15	max 999	max 250
	mean 1.42	mean 186.13	mean 1.42
	s.d. 2.25	s.d 254.31	s.d. 39.46

ALL WOOD SAWING ** ONE MINUTE

N=25

C.T. PITS	STRIAE	PITS	LINEAR GROOVES
NONE	min 0	min 0	min 0
	max 3	max 250	max 45
	mean .28	mean 42.52	mean 13
	s.d. .72	s.d. 56.71	s.d. 14.58

ALL WOOD SAWING ** 5 MINUTES

N=20

C.T. PITS	STRIAE	PITS	LINEAR GROOVES
NONE	min 0	min 5	min 0
	max 6	max 500	max 65
	mean 1.4	mean 118.7	mean 26.6
	s.d. 1.83	s.d 119.55	s.d 21.85

ALL WOOD SAWING ** 10 MINUTES

N=30

C.T. PITS	STRIAE	PITS	LINEAR GROOVES
NONE	min 0	min 0	min 0
	max 7	max 999	max 95
	mean 1.1	mean 151.17	mean 23.13
	s.d. 1.66	s.d. 212.53	s.d. 22.15

ALL WOOD SAWING ** 20 MINUTES

N=25

C.T. PITS	STRIAE	PITS	LINEAR GROOVES
NONE	min 0	min 10	min 0
	max 4	max 999	max 250
	mean 1.44	mean 206.4	mean 54.44
	s.d. 1.39	s.d. 239.27	s.d. 65.60

ALL WOOD SAWING ** 60 MINUTES

N=25

C.T. PITS	STRIAE	PITS	LINEAR GROOVES
NONE	min 0	min 45	min 3
	max 15	max 999	max 150
	mean 1.44	mean 405.36	mean 40.64
	s.d. 3.63	s.d. 350.81	s.d. 33.46

ALL WOOD SCRAPING ** TIMES COMBINED

N=75

C.T. PITS	STRIAE	PITS	LINEAR GROOVES
NONE	min 0	min 0	min 0
	max 8	max 500	max 80
	mean .84	mean 85.64	mean 19.07
	s.d. 1.73	s.d. 111.03	s.d. 19.64

ALL WOOD SCRAPING ** ONE MINUTE

N=15

C.T. PITS	STRIAE	PITS	LINEAR GROOVES
NONE	min 0	min 0	min 0
	max 8	max 400	max 45
	mean .6	mean 59	mean 8
	s.d. 1.99	s.d. 117.26	s.d. 14.34

ALL WOOD SCRAPING ** FIVE MINUTES

N=15

C.T. PITS	STRIAE	PITS	LINEAR GROOVES
NONE	min 0	min 10	min 0
	max 3	max 250	max 25
	mean .8	mean 50.47	mean 12
	s.d. 1.11	s.d. 59.67	s.d. 7.06

ALL WOOD SCRAPING ** TEN MINUTES

N=15

C.T. PITS	STRIAE	PITS	LINEAR GROOVES
NONE	min 0	min 0	min 0
	max 3	max 35	max 80
	mean .87	mean 94.67	s.d. 21.13
	s.d. 1.15	s.d. 114.06	s.d. 24.52

ALL WOOD SCRAPING ** TWENTY MINUTES

N=15

C.T. PITS	STRIAE	PITS	LINEAR GROOVES
NONE	min 0	min 8	min 0
	max 7	max 500	max 72
	mean 1.4	mean 108.73	mean 27.13
	s.d. 2.39	s.d. 119.78	s.d. 21.61

ALL WOOD SCRAPING ** SIXTY MINUTES

N=15

C.T. PITS	STRIAE	PITS	LINEAR GROOVES
NONE	min 0	min 30	min 5
	max 6	max 400	max 60
	mean .53	mean 115.33	mean 27.07
	s.d. 1.49	s.d. 116.28	s.d. 17.29

SCRAPING EUCALYPTUS ** ALL TIMES COMBINED

N=25

C.T. PITS	STRIAE	PITS	LINEAR GROOVES
NONE	min 0	min 0	min 0
	max 3	max 165	max 72
	mean .6	mean 48.84	mean 21.92
	s.d. 1.02	s.d. 43.06	s.d. 21.07

SCRAPING EUCALYPTUS ** ONE MINUTE

N=5

C.T. PITS	STRIAE	PITS	LINEAR GROOVES
NONE	NONE	min 0	NONE
		max 5	
		mean 1	
		s.d. 2	

SCRAPING EUCALYPTUS ** FIVE MINUTES

N=5

C.T. PITS	STRIAE	PITS	LINEAR GROOVES
NONE	min 0	min 10	min 0
	max 3	max 50	max 17
	mean 1.4	mean 25.4	mean 5.6
	s.d. 1.2	s.d. 14.44	s.d. 6.22

SCRAPING EUCALYPTUS ** TEN MINUTES

N=5

C.T. PITS	STRIAE	PITS	LINEAR GROOVES
NONE	min 0	min 33	min 11
	max 2	max 165	max 60
	mean .6	mean 84.6	mean 25.8
	s.d. .8	s.d. 46.16	s.d. 17.70

SCRAPING EUCALYPTUS ** TWENTY MINUTES

N=5

C.T. PITS	STRIAE	PITS	LINEAR GROOVES
NONE	min 0	min 37	min 26
	max 3	max 138	max 72
	mean 1	mean 81.6	mean 40
	s.d. 1.26	s.d. 35.58	s.d. 16.82

SCRAPING EUCALYPTUS ** SIXTY MINUTES

N=5

C.T. PITS	STRIAE	PITS	LINEAR GROOVES
NONE	NONE	min 30	min 12
		max 85	max 56
		mean 51.6	mean 38.2
		s.d. 21.58	s.d. 15.18

SCRAPING ACACIA ** ALL TIMES COMBINED

N=25

C.T. PITS	STRIAE	PITS	LINEAR GROOVES
NONE	min 0	min 0	min 0
	max 7	max 120	max 60
	mean .28	mean 28.4	mean 9.32
	s.d. 1.37	s.d. 32.75	s.d. 13.58

SCRAPING ACACIA ** ONE MINUTE

N=5

C.T. PITS	STRIAE	PITS	LINEAR GROOVES
NONE	NONE	min 0	min 0
		max 35	max 1
		mean 7	mean .2
		s.d. 14	s.d. .4

SCRAPING ACACIA ** FIVE MINUTES

N=5

C.T. PITS	STRIAE	PITS	LINEAR GROOVES
NONE	NONE	min 10	min 5
		max 50	max 15
		mean 27	mean 12
		s.d. 14	s.d. 3.69

SCRAPING ACACIA ** TEN MINUTES

N=5

C.T. PITS	STRIAE	PITS	LINEAR GROOVES
NONE	NONE	NONE	NONE

SCRAPING ACACIA ** TWENTY MINUTES

N=5

C.T. PITS	STRIAE	PITS	LINEAR GROOVES
NONE	min 0	min 8	min 0
	max 7	max 70	max 26
	mean 1.4	mean 29.6	mean 7.8
	s.d. 2.8	s.d. 21.13	s.d. 9.52

SCRAPING ACACIA ** SIXTY MINUTES

N=5

C.T. PITS	STRIAE	PITS	LINEAR GROOVES
NONE	NONE	min 45	min 7
		max 120	max 60
		mean 78.4	mean 26.6
		s.d. 27.49	s.d. 18.42

SCRAPING HARDWOOD ** ALL TIMES COMBINED

N=25

C.T. PITS	STRIAE	PITS	LINEAR GROOVES
NONE	min 0	min 25	min 5
	max 8	max 500	max 80
	mean 1.64	mean 179.68	mean 25.96
	s.d. 2.24	s.d. 143.46	s.d. 19.45

SCRAPING HARDWOOD ** ONE MINUTE

N=5

C.T. PITS	STRIAE	PITS	LINEAR GROOVES
NONE	min 0	min 40	min 7
	max 8	max 400	max 45
	mean 1.8	mean 169	mean 23.8
	s.d. 3.12	s.d. 151.27	s.d. 15.56

SCRAPING HARDWOOD ** 5 MINUTES

N=5

C.T. PITS	STRIAE	PITS	LINEAR GROOVES
NONE	min 0	min 25	min 14
	max 3	max 250	max 25
	mean 1	mean 99	mean 18.4
	s.d. 1.09	s.d. 82.12	s.d. 3.93

SCRAPING HARDWOOD ** 10 MINUTES

N=5

C.T. PITS	STRIAE	PITS	LINEAR GROOVES
NONE	min 0	min 55	min 10
	max 3	max 350	max 80
	mean 2	mean 199.4	mean 37.6
	s.d. 1.09	s.d. 129.87	s.d. 27.41

SCRAPING HARDWOOD ** 20 MINUTES

N=5

C.T. PITS	STRIAE	PITS	LINEAR GROOVES
NONE	min 0	min 60	min 5
	max 7	max 500	max 60
	mean 1.8	mean 215	mean 33.6
	s.d. 2.71	s.d. 151.79	s.d. 21.11

SCRAPING HARDWOOD ** 60 MINUTES

N=5

C.T. PITS	STRIAE	PITS	LINEAR GROOVES
NONE	min 0	min 40	min 5
	max 6	max 400	max 30
	mean 1.6	mean 216	mean 16.4
	s.d. 2.24	s.d. 154.22	s.d. 9.48

SAWING OAK ** ALL TIMES COMBINED

N=25

C.T. PITS	STRIAE	PITS	LINEAR GROOVE
NONE	min 0	min 0	min 0
	max 4	max 350	max 100
	mean 1.12	mean 102.84	mean 20.36
	s.d. 1.51	s.d. 103.75	s.d. 25.55

SAWING OAK ** ONE MINUTE

N=5

C.T. PITS	STRIAE	PITS	LINEAR GROOVES
NONE	NONE	min 0	min 0
		max 18	max 1
		mean 5	mean .2
		s.d. 23.40	s.d. .4

SAWING OAK ** 5 MINUTES

N=5

C.T. PITS	STRIAE	PITS	LINEAR GROOVES
NONE	NONE	min 5	min 2
		max 70	max 13
		mean 48.8	mean 8
		s.d. 23.40	s.d. 3.95

SAWING OAK ** 10 MINUTES

N=5

C.T. PITS	STRIAE	PITS	LINEAR GROOVES
NONE	min 0	min 12	min 0
	max 2	max 185	max 20
	mean .8	mean 97.4	mean 13
	s.d. .98	s.d. 68.17	s.d. 7.79

SAWING OAK ** 20 MINUTES

N=5

C.T. PITS	STRIAE	PITS	LINEAR GROOVES
NONE	min 0	min 20	min 0
	max 4	max 300	max 100
	mean 2	mean 143	mean 45.8
	s.d. 1.41	s.d. 113.82	s.d. 41.05

SAWING OAK ** 60 MINUTES

N=5

C.T. PITS	STRIAE	PITS	LINEAR GROOVES
NONE	min 0	min 130	min 27
	max 4	max 350	max 42
	mean 2.8	mean 220	mean 34.8
	s.d. 1.47	s.d. 88.09	s.d. 5.71

SAWING EUCALYPTUS ** TIMES COMBINED

N=45

C.T. PITS	STRIAE	PITS	LINEAR GROOVES
NONE	min 0	min 0	min 0
	max 15	max 999	max 250
	mean 1.38	mean 221.87	mean 37.4
	s.d. 2.51	s.d. 294.74	s.d. 46.65

SAWING EUCALYPTUS ** ONE MINUTE

N=5

C.T. PITS	STRIAE	PITS	LINEAR GROOVES
NONE	min 0	min 15	min 1
	max 2	max 250	max 45
	mean .4	mean 74	mean 16.4
	s.d. .8	s.d. 88.74	s.d. 15.56

SAWING EUCALYPTUS ** 5 MINUTES

N=10

C.T. PITS	STRIAE	PITS	LINEAR GROOVES
NONE	min 0	min 20	min 0
	max 4	max 350	max 65
	mean 1.2	mean 107	mean 27.2
	s.d. 1.4	s.d. 96.83	s.d. 22.04

SAWING EUCALYPTUS ** 10 MINUTES

N=10

C.T. PITS	STRIAE	PITS	LINEAR GROOVES
NONE	min 0	min 8	min 3
	max 2	max 200	max 65
	mean .8	mean 101.8	mean 24.4
	s.d. .87	s.d. 66.92	s.d. 22.01

SAWING EUCALYPTUS ** 20 MINUTES

N=10

C.T. PITS	STRIAE	PITS	LINEAR GROOVES
NONE	min 0	min 30	min 2
	max 3	max 999	max 250
	mean 1.3	mean 219	mean 57.4
	s.d. 1.35	s.d. 275.24	s.d. 72.85

SAWING EUCALYPTUS ** 60 MINUTES

N=10

C.T. PITS	STRIAE	PITS	LINEAR GROOVES
NONE	min 0	min 50	min 3
	max 15	max 999	max 150
	mean 2.7	mean 533.6	mean 51.1
	s.d. 4.58	s.d. 402.06	s.d. 48.09

SAWING ACACIA ** ALL TIMES COMBINED

N=25

C.T. PITS	STRIAE	PITS	LINEAR GROOVES
NONE	min 0	min 35	min 8
	max 10	max 999	max 200
	mean 2.12	mean 254.72	mean 50.2
	s.d. 2.53	s.d. 279.27	s.d. 42.25

SAWING ACACIA ** ONE MINUTE

N=5

C.T. PITS	STRIAE	PITS	LINEAR GROOVES
NONE	min 0	min 50	min 8
	max 3	max 150	max 45
	mean .6	mean 95	mean 30.6
	s.d. 1.2	s.d. 38.21	s.d. 13.84

SAWING ACACIA ** 5 MINUTES

N=5

C.T. PITS	STRIAE	PITS	LINEAR GROOVES
NONE	min 0	min 95	min 15
	max 6	max 300	max 65
	mean 3.2	mean 212	mean 44
	s.d. 2.04	s.d. 154.94	s.d. 16.55

SAWING ACACIA ** 10 MINUTES

N=5

C.T. PITS	STRIAE	PITS	LINEAR GROOVES
NONE	min 0	min 35	min 18
	max 5	max 75	max 42
	mean 1.2	mean 58	mean 27.4
	s.d. 1.94	s.d. 14.35	s.d. 8.94

SAWING ACACIA ** 20 MINUTES

N=5

C.T. PITS	STRIAE	PITS	LINEAR GROOVES
NONE	min 0	min 275	min 40
	max 4	max 800	max 200
	mean 1.4	mean 425	mean 108
	s.d. 1.49	s.d. 192.35	s.d. 60.46

SAWING ACACIA ** 60 MINUTES

N=5

C.T. PITS	STRIAE	PITS	LINEAR GROOVES
NONE	min 0	min 45	min 20
	max 10	max 999	max 75
	mean 4.2	mean 483.6	mean 41
	s.d. 3.37	s.d. 424.10	s.d. 18.81

SAWING PINE ** TIMES COMBINED

N=20

C.T. PITS	STRIAE	PITS	LINEAR GROOVES
NONE	min 0	min 0	min 0
	max 8	max 350	max 40
	mean 1	mean 76.7	mean 9.2
	s.d. 1.82	s.d. 114.16	s.d. 11.16

SAWING PINE ** ONE MINUTE

N=5

C.T. PITS	STRIAE	PITS	LINEAR GROOVES
NONE	min 0	min 0	min 2
	max 1	max 35	max 14
	mean .2	mean 16.6	mean 7
	s.d. .4	s.d. 13.62	s.d. 4.98

SAWING PINE ** 10 MINUTES

N=10

C.T. PITS	STRIAE	PITS	LINEAR GROOVES
NONE	min 0	min 0	min 0
	max 1	max 20	max 4
	mean .2	mean 8.2	mean 1
	s.d. .4	s.d. 8.06	s.d. 1.55

SAWING PINE ** 20 MINUTES

N=5

C.T. PITS	STRIAE	PITS	LINEAR GROOVES
NONE	min 0	min 10	min 0
	max 3	max 40	max 11
	mean 1.2	mean 26	mean 3.6
	s.d. 1.17	s.d. 10.68	s.d. 4.13

SAWING PINE ** 60 MINUTES

N=5

C.T. PITS	STRIAE	PITS	LINEAR GROOVES
NONE	min 0	min 80	min 11
	max 8	max 350	max 40
	mean 2.4	mean 256	mean 25.2
	s.d. 2.87	s.d. 93.51	s.d. 9.70

SAWING YEW ** ALL TIMES COMBINED

N=10

C.T. PITS	STRIAE	PITS	LINEAR GROOVES
NONE	min 0	min 0	min 0
	max 7	max 999	max 95
	mean 1.5	mean 280.9	mean 29.7
	s.d. 2.20	s.d. 319.01	s.d. 26.09

BONE WORKING ** ALL TIMES COMBINED

N=190

C.T. PITS	STRIAE	PITS	LINEAR GROOVES
min 0	min 0	min 0	min 0
max 300	max 60	max 300	max 47
mean 24.66	mean 3.91	mean 29.48	mean 6.15
s.d. 43.52	s.d.6.83	s.d. 38.38	s.d. 7.79

BONE WORKING ** ONE MINUTE

N=35

C.T. PITS	STRIAE	PITS	LINEAR GROOVES
min 0	min 0	min 0	min 0
max 140	max 15	max 30	max 7
mean 9.57	mean 1.09	mean 8.57	mean 1.29
s.d. 25.93	s.d. 2.91	s.d. 2.93	s.d. 1.85

ALL BONE WORK ** 5 MINUTES

N=30

C.T. PITS	STRIAE	PITS	LINEAR GROOVES
min 0	min 0	min 0	min 0
max 80	max 11	max 60	max 15
mean 8.57	mean 2.63	mean 16.87	mean 4.27
s.d. 16.35	s.d. 3.12	s.d. 16.39	s.d. 4.27

ALL BONE WORK ** 10 MINUTES

N=40

C.T. PITS	STRIAE	PITS	LINEAR GROOVES
min 0	min 0	min 0	min 0
max 300	max 60	max 150	max 22
mean 29.23	mean 7.4	mean 25.8	mean 5.33
s.d. 52.51	s.d. 12.24	s.d. 29.39	s.d. 5.56

ALL BONE WORKING ** 20 MINUTES

N=45

C.T. PITS	STRIAE	PITS	LINEAR GROOVES
min 0	min 0	min 0	min 0
max 200	max 15	max 100	max 23
mean 32.49	mean 3.13	mean 32.3	mean 6.29
s.d. 40.27	s.d. 3.36	s.d. 23.39	s.d. 5.88

ALL BONE WORKING ** 60 MINUTES

N=35

C.T. PITS	STRIAE	PITS	LINEAR GROOVES
min 0	min 0	min 4	min 0
max 300	max 19	max 250	max 40
mean 37.63	mean 5.09	mean 50.66	mean 12.11
s.d. 57.06	mean 4.95	s.d. 51.52	s.d. 10.79

BONE GRAVING ** TIMES COMBINED

N=45

C.T. PITS	STRIAE	PITS	LINEAR GROOVES
min 0	min 0	min 0	min 0
max 300	max 25	max 50	max 26
mean 38.71	mean 2.78	mean 10	mean 4.47
s.d. 67.73	s.d. 4.59	s.d. 9.65	s.d. 6.33

BONE SAWING ** TIMES COMBINED

N=85

C.T. PITS	STRIAE	PITS	LINEAR GROOVES
min 0	min 0	min 0	min 0
max 120	max 60	max 300	max 47
mean 16.55	mean 6.05	mean 43	mean 8.35
s.d. 23.34	s.d. 9.03	s.d. 49.26	s.d. 9.46

BONE GRAVING ** ONE MINUTE

N=10

min 0	min 0	min 0	min 0
max 140	max 2	max 17	max 7
mean 25.9	mean .3	mean 5.7	mean 2.4
s.d. 43.78	s.d. .64	s.d. 5.39	s.d. 4.27

BONE GRAVING ** 5 MINUTES

N=5

min 0	min 0	min 4	min 0
max 80	max 8	max 17	max 11
mean 21.4	mean 3	mean 10.6	mean 2.6
s.d. 30.39	s.d. 3.69	s.d. 5.39	s.d. 4.27

BONE GRAVING ** 10 MINUTES

N=10

min 0	min 0	min 0	min 0
amx 300	max 25	max 18	max 22
mean 56	mean 3.8	mean 7.7	mean 6.1
s.d. 91.45	s.d. 7.36	s.d. 5.57	s.d. 8.17

BONE GRAVING ** 20 MINUTES

N=10

C.T. PITS	STRIAE	PITS	LINEAR GROOVES
min 0	min 0	min 2	min 0
max 73	max 10	max 50	max 17
mean 16.5	mean 3.3	mean 13.5	mean 3.9
s.d. 24.06	s.d. 3.77	s.d. 13.63	s.d. 5.65

BONE GRAVING ** 60 MINUTES

N=10

C.T. PITS	STRIAE	PITS	LINEAR GROOVES
min 0	min 0	min 4	min 0
max 300	max 10	max 40	max 26
mean 65.1	mean 3.6	mean 12.8	mean 6.4
s.d. 86.73	s.d. 3.38	s.d. 11.19	s.d. 7.43

BONE SAWING ** TIMES COMBINED

N=85

C.T. PITS	STRIAE	PITS	LINEAR GROOVES
min 0	min 0	min 0	min 0
max 120	max 60	max 300	max 47
mean 16.55	mean 6.05	mean 43	mean 8.35
s.d. 23.34	s.d. 9.03	s.d. 49.26	s.d. 9.46

BONE SAWING ** ONE MINUTE

N=15

C.T. PITS	STRIAE	PITS	LINEAR GROOVES
min 0	min 0	min 0	min o
max 16	max 15	max 30	max 5
mean 2.3333	mean 2.33	mean 13.2	mean 1
s.d. 4.21	s.d. 4.09	s.d. 11.00	s.d. 1.59

BONE SAWING ** 5 MINUTES

N=15

C.T. PITS	STRIAE	PITS	LINEAR GROOVES
min 0	min 0	min 0	min 0
max 32	max 11	max 60	max 15
mean 4.4	mean 2.67	mean 25.6	mean 4.93
s.d. 8.65	s.d. 3.36	s.d. 18.69	s.d. 4.88

BONE SAWING ** 10 MINUTES

N=15

C.T. PITS	STRIAE	PITS	LINEAR GROOVES
min 0	min 0	min 6	min 0
max 66	max 60	max 150	max 16
mean 19	mean 15.06667	mean 40.66667	mean 5.53333
s.d. 22.44	s.d. 13.26	s.d. 37.95	s.d. 4.35

BONE SAWING ** 20 MINUTES

N=20

C.T. PITS	STRIAE	PITS	LINEAR GROOVES
min 1	min 0	min 12	min 1
max 120	max 15	max 100	max 23
mean 27.95	mean 3.9	mean 42.35	mean 8.95
s.d. 28.09	s.d. 3.58	s.d. 25.45	s.d. 6.28

BONE SAWING ** 60 MINUTES

N=15

C.T. PITS	STRIAE	PITS	LINEAR GROOVES
min 0	min 2	min 16	min 1
max 60	max 19	max 250	max 40
mean 21.33	mean 8.27	mean 72.07	mean 18
s.d. 21.24	s.d. 5.41	s.d. 65.05	s.d. 11.84

BONE SAWING ** 96 MINUTES

N=5

C.T. PITS	STRIAE	PITS	LINEAR GROOVES
min 0	min 1	min 25	min 8
max 80	max 4	max 300	max 47
mean 29	mean 2.2	mean 107	mean 17.8
s.d. 34.12	s.d. .98	s.d. 100.03	s.d. 14.66

BONE SCRAPING ** TIMES COMBINED

N=60

C.T. PITS	STRIAE	PITS	LINEAR GROOVES
min 0	min 0	min 0	min 0
max 200	max 9	max 100	max 23
mean 25.6	mean 1.73	mean 24.93	mean 4.3
s.d. 39.28	s.d. 2.01	s.d. 24.36	s.d. 4.83

BONE SCRAPING ** ONE MINUTE

N=10

C.T. PITS	STRIAE	PITS	LINEAR GROOVES
min 0	NONE	min 0	min 0
max 15		max 20	max 4
mean 4.1		mean 4.5	mean .6
s.d. 5.96		s.d. 6.41	s.d. 1.28

BONE SCRAPING ** 5 MINUTES

N=10

C.T. PITS	STRIAE	PITS	LINEAR GROOVES
min 0	min 0	min 0	min 0
max 26	max 6	max 20	max 8
mean 8.4	mean 2.4	mean 6.9	mean 2.8
s.d. 10.92	s.d. 2.33	s.d. 5.82	s.d. 2.48

BONE SCRAPING ** 10 MINUTES

N=15

C.T. PITS	STRIAE	PITS	LINEAR GROOVES
min 0	min 0	min 0	min 0
max 100	max 6	max 73	max 14
mean 21.6	mean 2.13	mean 23	mean 4.6
s.d. 25.16	s.d. 1.96	s.d. 19.89	s.d. 4.26

BONE SCRAPING ** 20 MINUTES

N=15

C.T. PITS	STRIAE	PITS	LINEAR GROOVES
min 0	min 0	min 8	min 0
max 200	max 9	max 60	max 14
mean 49.2	mean 2	mean 31.53	mean 4.33
s.d. 54.31	s.d. 2.28	s.d. 16.84	s.d. 3.59

BONE SCRAPING ** 60 MINUTES

N=10

C.T. PITS	STRIAE	PITS	LINEAR GROOVES
min 0	min 0	min 20	min 0
max 146	max 3	max 100	max 9
mean 34.9	mean 1.8	mean 56.4	mean 9
s.d. 45.01	s.d. 1.36	s.d. 25.77	s.d. 6.91

Appendix 4

Measurements on Striae Lengths For All Flint Tools

All measurements are in millimeters

ALL ANTLER WORKING ** DRY AND SOAKED COMBINED

N=45

Min	1
Max	60
Mean	10.91
S.d.	12.19

DRY ANTLER SAWING

N=33

Min	2
Max	60
Mean	12.18
S.d.	13.65

SOAKED ANTLER SAWING

N=12

Min	1
Max	19
Mean	7.42
S.d.	5.35

GRASS WORKING ** TILDEN PARK

N=2

Min	6
Max	10
Mean	8
S.d.	2

HIDE SCRAPING ** TANNED HIDE

N=4

Min	3
Max	5
Mean	3.75
S.d.	.83

ALL WOOD WORKING

N=111

Min 1
Max 50
Mean 7.71
S.d. 8.0

SAWING WOOD ** PINE

N=78

Min 1
Max 50
Mean 7.91
S.d. 9.2

SAWING WOOD ** EEUCALYPTUS

N=25

Min 2
Max 11
Mean 5.56
S.d. 2.4

SAWING WOOD ** OAK

N=8

Min 8
Max 15
Mean 12.5
S.d. 2.69

Appendix 5

Discriminant Analysis Summary and Tables

Discriminant Analysis Summary

Discriminant analysis was performed on the raw data counts for the major attributes found associated with use-wear on quartz. These attributes: comet-tailed pits, striae, linear grooves and pits were chosen because they were both identifiable and quantifiable on almost every experimental quartz tool. Comet-tailed pits, however, were <u>only</u> seen on those tools used for working either bone or antler.

These statistics were performed in an attempt to verify the subjective microscopic observations. Qualifying use-wear attributes is unavoidably subjective, as there exists no completely accurate (or easily implemented) method to measure features such as "brightness" and "luster", for example. Subjective terms such as "melted snow" are regularly used to describe certain optical qualities seen on polishes (Keeley 1980:56), being accepted by most researchers, but rejected by others as inadequate and unscientific terminology (refer to Cook and Dumont 1986 but also Dumont 1982, for a report on quantifying brightness). Still, these terms evoke strong images to most of us and, when viewing the polish in question, the descriptions seem appropriate and clear. Nonetheless, it is important to quantify those attributes, whenever possible, in order that we may chance upon a new way to objectively describe the microwear polishes on archaeological specimens clearly and accurately.

Discriminant analysis allowed grouping of the various polishes both by action and material worked, based on direct counts of the four attributes. Comet-tailed pits, striae, linear grooves, and pits were counted at 200X using the incident light microscope. At that magnification, the field of view is approximately 1mm X 1mm across and every attribute present in that "window" was counted. This count was repeated five times on each experimental tool. These are not random counts because, too often an area on a tool had to be discarded as the surface of the quartz was too irregular to permit focusing and, therefore, accurate counts could not be made. Some areas chosen had no visible attributes and, these were recorded as such. The sample was biased because I knew which areas on the experimental tools sustained the most use and probably tended to concentrate on those but, tried, nonetheless, to get as "good" a sample as possible over the entire worked edge.

In the test of discriminant analysis, the five observations per piece were added together giving one total attribute count per tool. Unfortunately, it is not possible to look at the totals of the attributes and determine what material a tool was used on and for how long.

As mentioned, discriminant analysis reclassifies the polishes based solely on the raw data counts without visual imput. It is then possible to see how the polishes are regrouped when sorted by action (graving, sawing, etc.), or material worked (bone, hide, etc.). Looking at the reclassification of materials that occurred after analysis, it is interesting that bone polish was reclassified in one instance as grass polish, and grass was reclassified once as a hide polish in the sixty minute table. When examining the tools microscopically, however, it would be highly unusual for a trained microwear specialist to mis-identify a bone polish as a grass polish or, to mistake a grass polish for a hide polish. The polish attributes of the three are quite distinct. The edge of a tool used for cutting grass becomes rounded and smooth, with a bright, metallic silica gloss. The bone polish, on the other hand, could be smooth and bright as well, but the presence of comet-tailed pits, would distinguish it immediately as a bone polish. Working hide produces nothing in the way of a smooth polish, rather, the edge of the tool becomes extremely rough and nubbly from continued attrition of the quartz surface scraping against the hide.

In the same vein, three quartz tools used as saws were reclassified as scrapers after sixty minutes of use, when the attributes were grouped into actions. Again, when examining the pieces microscopically, one would note the orientation of linear grooves and striae to the working edge and would be unlikely to mistake one for the other after an hour of use. These mis-classifications occur throughout both groupings and all time periods.

The results of this test are interesting but, disappointing if one hopes to use statistical methods to objectively identify use-wear polishes. At this time, there is no substitute comparable to the subjective identification of a polish based on its visual qualities and,

> ...it is probably naive to imagine that quantitative assessment will ever satisfactorily replace qualitative description based on observation, photomicrography and and the judgements of experienced analysts.
>
> (Cook and Dumont 1986 no pp.).

Multiple Discriminant Analysis: Classification by Action based on Four Attributes

60 minutes FROM ACTION	NUMBER OF OBSERVATIONS AND PERCENTS CLASSIFIED INTO ACTION CU	GR	SA	SC	TOTAL
CU	3 75.00	0 0.00	1 25.00	0 0.00	4 100.00
GR	0 0.00	2 66.67	0 0.00	1 33.33	3 100.00
SA	3 27.27	1 9.09	4 36.36	3 27.27	11 100.00
SC	5 45.45	3 27.27	1 9.09	2 18.18	11 100.00
TOTAL PERCENT	11 37.93	6 20.69	6 20.69	6 20.69	29 100.00

60 MINUTES FROM MATERIAL	NUMBER OF OBSERVATIONS AND PERCENTS CLASSIFIED INTO MATERIAL A	B	G	H	W	TOTAL
A	4 66.67	1 16.67	1 16.67	0 0.00	0 0.00	6 100.00
B	2 28.57	2 28.57	1 14.29	2 28.57	0 0.00	7 100.00
G	0 0.00	0 0.00	3 75.00	0 0.00	1 25.00	4 100.00
H	0 0.00	0 0.00	2 50.00	2 50.00	0 0.00	4 100.00
W	0 0.00	0 0.00	4 50.00	0 0.00	4 50.00	8 100.00
TOTAL PERCENT	6 20.69	3 10.34	11 37.93	4 13.79	5 17.24	29 100.00

Actions:	Materials:	Attributes:
CU = Cutting	A = Antler	Comet-tailed Pits
GR = Graving	B = Bone	Striae
SA = Sawing	G = Grass	Linear Grooves
SC = Scraping	H = Hide	Pits
	W = Wood	

Multiple Discriminant Analysis: Classification by Material based on Four Attributes

20 minutes FROM ACTION	NUMBER OF OBSERVATIONS AND PERCENTS CLASSIFIED INTO ACTION CU	GR	SA	SC	TOTAL
CU	5 100.00	0 0.00	0 0.00	0 0.00	5 100.00
GR	0 0.00	3 60.00	0 0.00	2 40.00	5 100.00
SA	2 16.67	3 25.00	3 25.00	4 33.33	12 100.00
SC	5 41.67	2 16.67	1 8.33	4 33.33	12 100.00
TOTAL PERCENT	12 35.29	8 23.53	4 11.76	10 29.41	34 100.00

20 MINUTES FROM MATERIAL	NUMBER OF OBSERVATIONS AND PERCENTS CLASSIFIED INTO MATERIAL A	B	G	H	W	TOTAL
A	2 28.57	2 28.57	1 14.29	2 28.57	0 0.00	7 100.00
B	0 0.00	7 77.78	0 0.00	2 22.22	0 0.00	9 100.00
G	0 0.00	0 0.00	5 100.00	0 0.00	0 0.00	5 100.00
H	0 0.00	0 0.00	3 60.00	2 40.00	0 0.00	5 100.00
W	0 0.00	0 0.00	3 37.50	1 12.50	4 50.00	8 100.00
TOTAL PERCENT	2 5.88	9 26.47	12 35.29	7 20.59	4 11.76	34 100.00

Actions:

CU = Cutting
GR = Graving
SA = Sawing
SC = Scraping

Materials:

A = Antler
B = Bone
G = Grass
H = Hide
W = Wood

Attributes:

Comet-tailed Pits
Striae
Linear Grooves
Pits

Multiple Discriminant Analysis: Classification of Action based on Four Attributes

10 MINUTES	NUMBER OF OBSERVATIONS AND PERCENTS CLASSIFIED INTO ACTION				
FROM ACTION	CU	GR	SA	SC	TOTAL
	3	0	0	1	4
CU	75.00	0.00	0.00	25.00	100.00
	1	3	0	1	5
GR	20.00	60.00	0.00	20.00	100.00
	3	2	4	4	13
SA	23.08	15.38	30.77	30.77	100.00
	6	1	2	3	12
SC	50.00	8.33	16.67	25.00	100.00
TOTAL	13	6	6	9	34
PERCENT	38.24	17.75	17.65	26.47	100.00

10 MINUTES	NUMBER OF OBSERVATIONS AND PERCENTS CLASSIFIED INTO MATERIAL					
FROM MATERIAL	A	B	G	H	W	TOTAL
	5	0	1	1	1	8
A	62.50	0.00	12.50	12.50	12.50	100.00
	1	4	1	2	0	8
B	12.50	50.00	12.50	25.00	0.00	100.00
	0	0	4	0	0	4
G	0.00	0.00	100.00	0.00	0.00	100.00
	0	0	1	4	0	5
H	0.00	0.00	20.00	80.00	0.00	100.00
	0	0	5	0	4	9
W	0.00	0.00	55.56	0.00	44.44	100.00
TOTAL	6	4	12	7	5	34
PERCENT	17.65	11.76	35.29	20.59	14.71	100.00

Actions:	Materials:	Attributes:
CU = Cutting	A = Antler	Comet-tailed Pits
GR = Graving	B = Bone	Striae
SA = Sawing	G = Grass	Linear Grooves
SC = Scraping	H = Hide	Pits
	W = Wood	

Multiple Discriminant Analysis: Classification by Material based on Four Attributes

5 MINUTES FROM ACTION	NUMBER OF OBSERVATIONS AND PERCENTS CLASSIFIED INTO ACTION CU	GR	SA	SC	TOTAL
CU	4 100.00	0 0.00	0 0.00	0 0.00	4 100.00
GR	0 0.00	3 75.00	1 25.00	0 0.00	4 100.00
SA	2 18.18	1 9.09	4 36.36	4 36.36	11 100.00
SC	6 50.00	2 16.67	1 8.33	3 25.00	12 100.00
TOTAL PERCENT	12 38.71	6 19.35	6 19.35	7 22.58	31 100.00

5 MINUTES FROM MATERIAL	NUMBER OF OBSERVATIONS AND PERCENTS CLASSIFIED INTO MATERIAL A	B	G	H	W	TOTAL
A	6 66.67	2 22.22	1 11.11	0 0.00	0 0.00	9 100.00
B	2 33.33	2 33.33	1 16.67	1 16.67	0 0.00	6 100.00
G	0 0.00	0 0.00	4 100.00	0 0.00	0 0.00	4 100.00
H	0 0.00	0 0.00	2 40.00	3 60.00	0 0.00	5 100.00
W	0 0.00	0 0.00	2 28.57	2 28.57	3 42.86	7 100.00
TOTAL PERCENT	8 28.51	4 12.90	10 32.26	6 19.35	3 9.68	31 100.00

Actions:

CU = Cutting
GR = Graving
SA = Sawing
SC = Scraping

Materials:

A = Antler
B = Bone
G = Grass
H = Hide
W = Wood

Attributes:

Comet-tailed Pits
Striae
Linear Grooves
Pits

Multiple Discriminant Analysis: Classification by Action based on Four Attributes

1 MINUTE	NUMBER OF OBSERVATIONS AND PERCENTS CLASSIFIED INTO ACTION				
FROM ACTION	CU	GR	SA	SC	TOTAL
	4	0	0	0	4
CU	100.00	0.00	0.00	0.00	100.00
	2	2	1	0	5
GR	40.00	40.00	20.00	0.00	100.00
	4	0	7	1	12
SA	33.33	0.00	58.33	8.33	100.00
	7	0	1	4	12
SC	58.33	0.00	8.33	33.33	100.00
TOTAL	17	2	9	5	33
PERCENT	51.52	6.06	27.27	15.15	100.00

1 MINUTE	NUMBER OF OBSERVATIONS AND PERCENTS CLASSIFIED INTO MATERIAL					
FROM MATERIAL	A	B	G	H	W	TOTAL
	5	0	4	0	0	9
A	55.56	0.00	44.44	0.00	0.00	100.00
	0	4	3	0	0	7
B	0.00	57.14	42.86	0.00	0.00	100.00
	0	0	4	0	0	4
G	0.00	0.00	100.00	0.00	0.00	100.00
	0	0	2	1	2	5
H	0.00	0.00	40.00	20.00	40.00	100.00
	0	0	4	0	4	8
W	0.00	0.00	50.00	0.00	50.00	100.00
TOTAL	5	4	17	1	6	33
PERCENT	15.15	12.12	51.52	3.03	18.18	100.00

Actions:	Materials:	Attributes:
CU = Cutting	A = Antler	Comet-tailed Pits
GR = Graving	B = Bone	Striae
SA = Sawing	G = Grass	Linear Grooves
SC = Scraping	H = Hide	Pits
	W = Wood	

Bibliography

Anderson-Gerfaud, P.

1983. A consideration of the uses of certain backed and "lustred" stone tools from Late Mesolithic and Natufian levels of Abu Hureyra and Mureybet (Syria). In Traces de l'Utilisation sur les Outils Neolithiques du Proche Orient (ed. M.-C. Cauvin):77-106. Lyon:Maison de l'Orient Mediterraneen.

Anderson, P.

1980. A testimony of prehistoric tasks: diagnostic residues on stone tool working edges. World Archaeology 12(2): 181-194.

Andrews, J.T.

1975. An Approach to Glaciers and Their Environments. Massachusetts, Duxbury Press.

Balout, L.

1955. Prehistoire de l'Afrique du Nord: Essai de Chronologie. Paris.

Beaumont, P.B.

1973. "Border Cave - a progress report," South African Journal of Science 69:41-46.

Beaumont, P.B., deVillier K. and J.V. Vogel

1978. Modern man in sub-saharan Africa prior to 49,000 B.P.: a review and evaluation with particular reference to Border Cave. South African Journal of Science 74:409-419.

Beck, B.B.

1980. Animal Tool Behavior. New York: Garland STPM Press.

Beilby, G.T.

1904. Surface flow in crystalline solids under mechanical disturbance. Proceedings of the Royal Society of London 72:218-225.

Bettison, C.A.

1985. An experimental approach to sickle sheen deposition and archaeological interpretation. Lithic Technology 14(1):26-32.

Beyries, S.

1982. Comparaison de traces d'utilisation sur differentes roches siliceuses. In Recent Progress in Microwear Studies. Praehistorica Belgica #2 (ed. D. Cahen):235-240 Tervuren: Musee Royale de l'Afrique Centrale.

Binford, L.

1981. Bones. Ancient Men and Modern Myths. New York:Academic Press.

1973. Interassemblage variability - the Mousterian and the "functional" argument. In The Explanation of Culture Change (ed. C. Renfrew):227-254. London:Duckworth.

Binford, L.R. and Binford, S.

1966. A preliminary analysis of functional variability in the Mousterian of Levallois facies. In Recent Studies in Paleoanthropology, J.D. Clark and F.C.Howell, eds., American Anthropology 68:2 Part 2,238-295.

Binford, S.

1968. Ethnographic data and understanding the Pleistocene. In Man the Hunter. (R.B.Lee and I.Devore eds.) Chicago:Aldine-Atherton 274-275.

Boesch, C. and Boesch, H.

1981. Sex differences in the use of natural hammers by wild chimpanzees: a preliminary report. Journal of Human Evolution 10:585-593.

1983. Optimization of nut-cracking with natural hammer by wild chimpanzees. Behavior 83:265-286.

Bordes, F.

1950. Principles d'une methode d'etude des techniques de debitage et de la typologie de Paleolithique ancien et moyen. L'Anthropologie. 54:19-34.

1961. Typologie du Palaeolithique Ancien et Moyen. Publications de l'Universite de Bordeaux Memoir No.1.

1968. The Old Stone Age. New York: World University Library, McGraw Hill Co.

Bordes, F. and Crabtree, D.

1969. The Corbiac blade technique and other experiments. Tebiwa. 12(2):1-21.

Bordes, F. and de Sonneville-Bordes, D.

1970. The significance of variability in Paleolithic assemblages _World Archaeology_. 2:61-73.

Bourlon, M. and Bourlon, J. and Bouyssonie, A.

1912. Grattoirs carenes; rabots et grattoirs nucleiformes. Essa de classification des grattoirs. _Revue Anthropologique_. 22:473-486.

Braidwood, R.

1946. Terminology in Prehistory. In _Human Origins: An Introductory General Course in Anthropology: Selected Readings Series II_. Chicago: University of Chicago Press.

Brink, J.

1978. The role of abrasives in the formation of lithic use-wear. _Journal of Archaeological Science_ 5:363-371.

Bunn, H.T.

1981. Archaeological evidence for meat-eating by Plio-Pleistocene hominids from Koobi Fora and Olduvai Gorge. _Nature_. 291:574-577.

Bunn, H.T.,Harris, J.W.K.,Isaac, G.,Kaufulu, Z.,Kroll, E.,Schick, K. Toth, N, and Behrensmeyer, A.K.

1980. FxJj 50: An early Pleistocene site in northern Kenya. _World Archaeology_. 12:109-136.

Burkitt, M.

1963. _The Old Stone Age._ New York: Atheneum.

Cahen, D. and Keeley, L.

1980. Not less than two, not more than three. _World Archaeology_ 12(2):166-180.

Cahen, D., Keeley, L.H. and Van Noten, F.L.

1979. Stone tools, toolkits and human behavior in prehistory. _Current Anthropology_. 20: 661-683.

Childe, G.

1925. _The Danube in prehistory. The Dawn of European Civilization_.

Clark, G.

1969. *World Prehistory A New Outline*. Cambridge University Press.

1970. *Aspects of Prehistory*. University of California Press.

Clark, J.D.

1950. The newly discovered Natchikufu culture of Northern Rhodesia. *South African Archaeological Bulletin* 5(19):86-98.

1958 Some stone age woodworking tools in southern Africa. *South African Archaeological Bulletin* 13(52):144-152.

1967 Problems of archaeological nomenclature and definition in the Congo Basin. In *Sixieme Congres Panafricain de Prehistoire. Dakar* (ed. H. Hugot): 584-593.

1968. Studies of hunter-gatherers as an aid to the interpretation of prehistoric societies. In *Man the Hunter*. (R.B. Lee and I. Devore, eds). Chicago: Aldine-Atherton.

1970. *The Prehistory of Africa*. New York: Praeger Publishers.

1980 Raw material and African lithic technology. *Man and his Environment* 14:44-55.

Clark, J.D. and Haynes, C.V.

1970. An elephant butchery site at Mwanganda's village, Karonga, Malawi, and its relevance for Palaeolithic archaeology. *World Archaeology*. 1:390-411.

Clark, J.D. and Kleindienst, M.R.

1974. The stone age cultural sequence: terminology, typology and raw material. Chapter 4. In *Kalambo Falls Prehistoric Site. Vol. II*. (J.D. Clark, ed.) Cambridge: Cambridge University Press. Pp 71-103.

Clark, J.D. and Prince, G.R.

1978 Use-wear on later stone age microliths from Laga Oda Haraghi, Ethiopia and possible functional interpretations. *Azania* 13:101-110.

Clark, W.E.L.

1962. *History of the Primates*. 8th edition. London:Trustees of the British Museum 75-76.

Close, A.

1978. The identification of style in lithic artifacts. *World Archaeology*. 10(3): 223-237.

Cook, J. and Dumont, J.

1986. The development and application of microwear analysis since 1964. In *The Human Uses of Flint and Chert - Papers from the 4th International Flint Symposium* (eds. G. Sieveking and M. Newcomer). Cambridge: Cambridge University Press.

Crabtree, D.

1972. *An Introduction to Flintworking*. Occasional Papers of the Idaho State University Museum, Number 28. Pocatello, Idaho.

Curwin, E.C.

1930. Prehistoric flint sickles. *Antiquity*, 4:179-180.

Daniel, G.

1975. *150 Years of Archaeology*. London: Duckworth.

Daniels, S.G.H.

1967. The genesis of types:a logical model. In *Sixieme Congres Panafricain de Preshistoire, Dakar*.

Del Bene, T.

1979. Once upon a striation:current models of striation and polish formation. In *Lithic Use-Wear Conference*. (B. Hayden, ed). New York: Academic Press.

Diamond, G.

1979. The nature of so-called polished surfaces on stone artefacts. In *Lithic Use-Wear Analysis*. (B. Hayden, ed). New York: Academic Press.

Dumont, J.V. and Cook, J.

1983. The development and application of microwear analysis since 1960. Paper presented to the Fourth International Flint Symposium. Brighton, England.

1982 The quantification of microwear traces: a new use for interferometry. *World Archaeology* 14(2):206-217.

Dunnell, R.C.

1978. Style and function: a fundamental dichotomy. _American Antiquity_. 43(2):192-202.

Fedje, D.

1979. S.E.M. analysis of use-striae. In _Lithic Use-Wear Analysis_. (B. Hayden, ed.). New York: Academic Press.

Fennenga, G.

1982. Unpublished research paper, University of California, Berkeley, Department of Anthropology.

Flenniken, J. and Haggarty, J.

1979. Trampling as agency in the formation of edge damage: an experiment in lithic technology. _Northwest Anthropology Research Notes_ 13(2):208-214.

Flood, J.

1983. _Archaeology of the Dreamtime_. Honolulu:University of Hawaii Press.

Fojud, R. and Kobiewicz, M.

1982. The influence of soluble salts on the state of preservation of flint surfaces and use-wear traces. In _Recent Progress in Microwear Studies_ (ed. D.Cahen): 245-250. Tervuren: Musee Royale de l'Afrique Centrale.

Francis, M.E.

1912. _The Book of Grasses_. New York: Doubleday, Page and Co.

Gero, J.M.

1978. Summary experiments to duplicate post excavational damage to tool edges. _Lithic Technology_, 7:34.

Goodall, J.

1964. Tool-using and aimed throwing in a community of free-living chimpanzees. _Nature_, 201:1264-1266.

1968. The behaviour of free-living chimpanzees in the Gombe Stream Reserve. _Animal behavior Monograph_ 1:161-311.

Goodwin, A.J.H.

1958. Formative years of our prehistoric terminology, _South African Archaeological Bulletin_ 13(49):25-33.

Gould, F.W.

1968. *Grass Systematics*. New York: McGraw-Hill Book Co.:169.

Gould, R., Koster, D.A. and Sontz, A.H.

1971. The lithic assemblages of the Western Desert Aborigines of Australia. *American Antiquity*, 36:146-169.

Grace, R., Graham, I.D. and Newcomer, M.H.

1985. The quantification of microwear polishes. *World Archaeology* 17:112-120.

Guilmet, G.

1977. The evolution of tool-using and tool making behavior. *Man* 12:33-47.

Hay, C.A.

1977. Use-scratch morphology: a functionally significant aspect of edge damage on obsidian tools. *Journal of Field Archaeology*. 4:491-494.

Isaac, G.

1968. Traces of Pleistocene hunters: an East African Example. In *Man the Hunter*. (R.B. Lee and I.Devore eds.) Chicago: Aldine-Atherton.

1971. The diet of early man: aspects of archaeological evidence from lower and middle Pleistocene sites in Africa. *World Archaeology* 1:1-28.

1978. The food sharing behavior of protohuman hominids. *Scientific Americs*. 238(4): 90-106.

Jeffreys, M.D.W.

1951. Quartz hand-axes. *South African Archaeological Bulletin*. 6:32-33.

Jones, P.

1979. Effects of raw material on biface manufacture. *Science*. 204:835-836.

1980. Experimental butchery with modern stone tools and its relevance for Paleolithic archaeology. *World Archaeology* 12(2):153-165.

Kamminga, J.

1977. A functional study of use-polished elouras. In Stone Tools as Cultural Markers (R.V.S.Wright, ed.): 205-212. Canberra: Australian Institute of Aboriginal Studies.

1978. A Journey into the Microcosms. A Functional Analysis of Certain Classes of Prehistoric Australian Stone Tools. Unpublished Ph.D. dissertation University of Sydney.

1979. The nature of use-polish on prehistoric stone tools. In Lithic Use-Wear Analysis. (B. Hayden, ed.):143-157.

Kantman, S.

1970. Equisse d'un procede analytiques pour l'etude macrographiques des "encoches". Quaternaria.13:269-280.

1970. Raclettes mousteriennes: une etude experimental sur le distinction de retouches intentionelle et les modifications du tranchant par utilisation. Quaternaria 13:295-304.

Keeley, L.H.

1974. Technology and methodology in microwear studies:a critical review. World Archaeology 5:323-336.

1977. The functions of palaeolithic flint tools. Scientific American 237(5):109-126.

1980. Experimental Determination of Stone Tool Uses. A Microwear Analysis. Chicago: The University of Chicago Press.

Keeley, L.H. and Newcomer, M.

1977. Microwear analysis of experimental flint tools: a test case.Journal of Archaeological Science 4:29-62.

Keeley, L.H. and Toth, N.

1981. Microwear polishes on early stone tools from Koobi Fora, Kenya. Science 293: 464-465

Kleindienst, M.

1961. Variability within the Late Acheulian assemblage in eastern Africa. South African Archaeological Bulletin 16:35-52.

1967. Questions of terminology in regard to the study of Stone Age industries in eastern Africa -culture stratigraphic units. In Background to Evolution in Africa. (W.W. Bishop and J.D. Clark, eds.) Chicago: 821-859.

Lanpo, J.

1980. Early Man in China, Beijing: Foreign Language Press.

Laville, H., J.P. Rigaud and J. Sackett

1980. Rock Shelters of the Perigord. New York: Academic Press.

Lawn, B.R. and D.B. Marshall

1979. Mechanism of microcontact fracture in brittle solids In Lithic Use-Wear Analysis. (B. Hayden ed.) New York: Academic Press.

Leakey, L.S.B.

1950a. Terminology in prehistory. South African Archaeological Bulletin. 5(17):20-22.

1950b. Stone implements: how they were made and used. South African Archaeological Bulletin. 18:71-74.

Leakey, M.D.

1970. Stone artefacts from Swartkrans. Nature. 225(5238):1124-1127.

1971. Olduvai Gorge Volume 3: Excavations in Beds I and II, 1960-1963. Cambridge: Cambridge University Press.

Levi-Sala, I.

1986. Use-wear and post-depositional surface modification: a word of caution. Journal of Archaeological Science 13:229-244.

Libby, W.F., Anderson, E.C. and Arnold, J.R.

1949. Age determination by radiocarbon content: worldwide assay of natural radiocarbon. Science. 109(2827):227-228.

Lovejoy, C.O.

1981. The origin of man. Science 211(4480):341-350.

Mansur, M.E.

1983. Scanning electron microscopy of dry hide working tools:the role of abrasives and humidity in microwear polish. _Journal of Archaeological Science_ 10(3):223-230.

Mason, R.J.

1962. The Sterkfontein artefacts and their maker. _South African Archaeological Bulletin_. 17(66):109-125.

1967. Cultures, technology and analytical procedures in African prehistory. In _Sixieme Congres Panafricain de Prehistoire, Dakar_.

McGrew, W.C., Tutin, C.E.G. and Baldwin, P.J.

1979. Chimpanzees, tools and termites: cross-cultural comparisons of Senegal, Tanzania and Rio Muni. _Man_ 14:185-214.

Meeks, N.D., G. de G. Sieveking, M.S. Tite and J. Cook

1982. Gloss and use-wear traces on flint sickles and similar phenomena. _Journal of Archaeological Science_ 9:317-340.

Mears, B.

1978. _The Changing Earth: Introduction to Geology_. New York: D. Van Nostrand Co.

Mellars, P.

1970. Some comments on the notion of 'functional variability' in Palaeolithic assembalges. _World Archaeology_. 14(1):26-32.

Merrick, H.V.

1976. Recent archaeological research in the Plio-Pleistocene deposits of the lower Omo Valley, southwestern Ethiopia. In _Human Origins. Louis Leakey and the East African Evidence_ (G. Ll. Isaac and E.R. McCown, eds.):461-482. Menlo Park:W.A. Benjamin Inc.

Merrick, H.V. and Merrick, J.P.S.

1976. Archaeological occurrences of earlier Pleistocene age, from the Shungura Formation. In _Earliest Man and Environments in the Lake Rudolf Basin_ (Y. Coppens et al, eds.):574-584. Chicago: University of Chicago Press.

Merrick, H.V., de Heinzelin, J., Haesaerts, P. and Howell, F.C.

1973. Archaeological occurrences of early Pleistocene age from the Shungura Formation, Lower Omo Valley, Ethiopia. _Nature_, 242: 572-575.

Miller, S.

1969. _The Natchikufan Industries of the Later Stone Age in Zambia_. Doctoral Dissertation in Anthropology, University of California, Berkeley.

Moss, E.

1983a. The functional burins and tanged points from Tell Abu Hureyra (Syria). In _Traces d'Utilisation sur les Outils Neolithiques de Proche Orient_ (M.-C.Cauvin, ed):143-161. Lyon: Maison de l'Orient Mediterraneen.

1983b. _The Functional Analysis of Flint Implements_ Oxford: BAR International Series 177.

1983c. Some comments on edge damage as a factor in functional analysis of stone artifacts. _Journal of Archeological Science_ 10:231-242.

Movius, H.L.

1948. The lower Palaeolithic cultures of southern and eastern Asia. _Transactions of the American Philosophical Society_. Vol 38:4.

Newcomer, M.H.

1976. Spontaneous Retouch. _2nd International Symposium on Flint. Staringia_ 3:62-64.

Newcomer, M.H., Grace, R. and Unger-Hamilton, R.

1986. Investigating microwear polishes with blind tests. _Journal of Archaeological SCience_. 13:203-217.

Nishida, T. and Vehara, S.

1980. Chimpanzees, tools and termites: another example from Tanzania. _Current Anthropology_ 21:671-672.

Nisson, K. and Dittemore, M.

1974. Data and wear pattern analysis: a study of socketed Eskimo scrapers. _Tebiwa_, 17(1).

Oakley, K.

1956. _Man the Toolmaker_. 3rd edition London: British Museum of Natural History.

Odell, G.H.

1983. Problemes dans l'etude des traces d'utilisation. In _Traces d'Utilisation sur les Outils Neolithiques de Proche Orient_ (ed. M.-C.Cauvin):17-24. Lyon: Maison de l'Orient Mediterraneen.

1981. The morphological express at function junction. Searching for meaning in lithic tool types. _Journal of Anthropological Research_. 37:319-342.

1980. Butchering with stone tools: some experimental results. _Lithic Technology_, 9(2):39-48.

1975. Microwear in perspective: a sympathetic response to Lawrence H.Keeley. _World Archaeology_ 7:226-240.

Odell, G.H. and Odell-Vereecken, F.

1980. Verifying the reliability of lithic use-wear assessments by "blind tests" the low power approach. _Journal of Field Archaeology_ 7(1):87-120.

Paterson, W.S.B.

1969. _The Physics of Glaciers_. London, The Pergamon Press.

Phillipson, L. and Phillipson, D.

1970. Patterns of edge damage on the Late stone-age industry from Chiwemupula, Zambia. _Zambia Museums Journal_. 1:40-75.

Piggott, S.

1973. _Ancient Europe_. Fourth Printing. Chicago:Aldine Publishing Co.

Plisson, H.

1982. Une analyse fonctionelle des outillages basaltiques. In _Praehistorica Belgica #2_ (D.Cahen):235-244. Tervuren: Musee Royale de l'Afrique Centrale.

Plisson, H. and Mauger, H.

1987. Chemical and mechanical alteration of microwear polishes: an experimental approach. In _The Human Uses of Flint and Chert_ Papers presented at the 4th International Flint Symposium, Brighton, England. April 1983. (G. Sieveking and M. Newcomer, eds). Cambridge: Cambridge University Press.

Pope, S.T.

1918. *California Publication in American Archaeology and Ethnology*. 13:3.

1974. Hunting with Ishi. *Journal of California Anthropology* 1:2.

Potts, R. and P. Shipman

1981. Cutmarks made by stone tools on bones from Olduvai Gorge, Tanzania. *Science* 291:577-580.

Rabinowicz, E.

1968. Polishing. *Scientific American* 218:99-109.

Read, H.H. and Watson, J.

1966. *Beginning Geology*. New York: St. Martin's Press.

Rigaud, J.P.

1978. The significance of variability among lithic artifacts: a specific case from southwestern France. *Journal of Anthropological Research*. 34:299-310.

Sackett, J.

1966. Quantitative analysis of upper Palaeolithic stone tools. In *Recent Studies in Palaeoanthropology*. (J.D. Clark and F.C. Howell, eds.). Special Publications of *American Anthropologist* 68(2) Part 2: 356-394.

1973. Style, function, and artifact variability in Palaeolithic assemblages. In *The Explanation of Culture Change* (C. Renfrew, ed.): 317-325. London: Duckworth.

1981. From de Mortillet to Bordes: a century of French Palaeolithic research. In *Towards a History of Archaeology* (G. Daniel, ed.):85-99. London: Thames and Hudson.

1982. Approaches to style in lithic archaeology. *Journal of Anthropological Archaeology*, 1:59-112.

Sanders, J.E.; Anderson,A. H, Jr. and Carola, R.

1976. *Physical Geology*. New York, Harper and Row.

Semenov, S.A.

1964. *Prehistoric Technology*. Translated by M. Thompson. London: Cory, Adams and Mackay.

Shackley, M.

1974. Stream abrasion of flint implements. Nature 248(5448):501-502.

Shipman, P. and Rose, J.

1983. Early hominid hunting, butchering, and carcass-processing behaviors: approaches to the fossil record. Journal of Anthropological Archaeology. 2:57-98.

Shipman, P., Potts, R. and Pickford, M.

1983. Lainyamok, a new Middle Pleistocene hominid site. Nature. 306(5941): 365-368.

Silberbauer, G. B.

1981. Hunter and habitat in the central Kalahari Desert. Cambridge: Cambridge University Press.

Smith, P.E.

1982. The late Palaeolithic and Epi-Palaeolithic of northern Africa. In Cambridge History of Africa (J.D. Fage and R. Oliver, eds.): 342-409. Cambridge: Cambridge University Press.

Speck, F.G.

1937. Analysis of Eskimo and Indian skin-dressing methods in Labrador. Ethnos. 2(6):345-353.

Spurrell, F.C.J.

1892. Notes on early sickles. Archaeological Journal 49: 53-60.

Sugiyama, Y. and Koman, J.

1979. Tool-using and making behavior in wild chimpanzees at Bossou, Guinea. Primates 20:513-524.

Summers, R.

1950. Armchair archaeology. South African Archaeological Bulletin 5(19):101-103.

1967. An experimental application of the Wartenstein recommendations for the archaeological terminology and nomenclature. In Sixieme Congres Panafricain du Prehistoire, Dakar.

Susman, R.

1980. Acrobatic pygmy chimpanzees. Natural History 89:32-39.

Susman, R.L., Stern, J.T. and Jungers, W.L.

1985. Locomotor adaptations in the Hadar hominids. In Ancestors: The Hard Evidence. (E. Delson, ed.). New York: Alan R. Liss.

Sussman, C.

1985a. Microwear Analysis of Experimental Quartz Tools. Anthroquest 32:7-8. The L.S.B. leakey Foundation Newsletter.

1985b. A Preliminary Microwear Study of Select Tools from Olduvai Gorge, Tanzania. Progress report made to the L.S.B. Leakey Foundation.

1985c. Microwear on quartz: fact or fiction? World Archaeology. 17:101-111.

Symens, N.

1986. A Functional Analysis of Selected Stone Artefacts from the Magdalenian Site of Verberie, France. Journal of Field Archaeology 13:213-222.

Tindale, N.B.

1965. Stone implement making among the Nakako Nagadajara and Pitjanjara of the Great Western Desert. Records of the South Australian Museum. 15(1).

Tixier, J.

1963. Typologie de L'Epipalaeolithique du Maghreb. Vol. II. Memoires du Centre de Recherches Anthropologiques Prehistoriques et Ethnographiques. Paris:Alger.

1967. Procedes d'analyse et questions de terminologie concernant l'etude des ensembles indutriels du Paleolithique et de l'Epipaleolithique dans l'Afrique du nord-ouest. In Background to Evolution in Africa. (W.W. Bishop and J.D. Clark, eds.). Chicago: University of Chicago Press.

1972. Obtention de lames pa debitage "sous le pied". Bulletin de la Societe Prehistorique Francais 69:134-139.

Tixier, J., Inizan, M.L. and Roche, H.

1980. Prehistoire de la pierre taillee. 1:terminologie et technologie. Antibes: Cercle de Recherches et d'Etudes Prehistoriques.

Tomenchuck, J.

1983. Predicting the past: examples from the use wear study of selected chipped stone tools from two Epipalaeolithic occupations in Israel. In Traces d'Utilisation sur led Outils Neolithiques du Proche Orient. (M.-C. Cauvin, ed.):57-76. CNRS Lyon: Maison de l'Orient Mediterraneen.

Toth, N.

1982. The Stone Technologies of Early Hominids at Koobi Fora, Kenya: An Exoerimental Approach. Ph.D. thesis. Berkeley: University of California at Berkeley, Department of Anthropology.

Tringham, R.; Cooper, G.; Odell, G.; Voytek, B.; and Whitman, A.

1974. Experimentation in the formation of edge damage: a new approach to lithics. Journal of Field Archaeology 1:171-196.

Trustees of the British Museum

1968. Flint Implements. An Account of Stone Age Techniques and Cultures. William Clowes and Sons.

Tucker, G.C.

1980. Quantitative affirmation of intuitive typology. Tebiwa. 22:1-11.

Unger-Hamilton, R.

1984. The formation of use wear polish on flint: beyond the "deposit versus abrasion" controversy. Journal of Archaeological Science, 11:91-98.

1983. An investigation into the variables affecting the development and appearance of plant polish on flint blades. In Traces D'Utilisation sur les Outils Neolithiques du Proche Orient (M.-C. Cauvin, ed.): 243-250. Lyon: Maison de l'Orient Mediterraneen.

Vaughan, P.

1981. Microwear analysis of experimental flint and obsidian tools. Stariniga 6:90-91.

1983. La fonction des outils prehistoriques. La Recherche. 14(148):1226-1234.

Vial, A.E.L.

1952. Alpine Glaciers. London: The Batchworth Press.

Vincent, A.S.

1985. Wild Tubers as a Harvestable resource in the East African Savannas: Ecological and Ethnographic Studies. Unpublished Pd.D. Dissertation. Berkeley: University of California, Department of Anthropology.

Wheat, M.

1967. Survival Arts of the Primitive Paiutes. Nevada: University of Nevada Press.

Witthoft, J.

1967. Glazed polish on flint tools. American Antiquity. 32:282-388.

Wood, B.

1976. The Evolution of Early Man. Cassell Australia Ltd.

Yamada, S. and Shimura, M.

1984. Use-wear analysis of stone tools with scanning electron microscope. In Microbeam Analysis (A.D. Romig, Jr. and J.I. Goldstein, eds.):223-226. San Francisco: San Francisco Press.

Yellen, J.

1977. Archaeological Approaches to the Present. New York: Academic Press.

Zihlman, A.

1979. Pygmy chimpanzee morphology and the interpretation of early hominids. South African Journal of Science 75(4):

Zihlman, A., Cronin, J.E., Cramer, D.L. and Sarich, V.

1978. Pygmy chimpanzee as a possible prototype of the common ancestor of humans, chimpanzees and gorillas. Nature 275:744-746.

www.ingramcontent.com/pod-product-compliance
Lightning Source LLC
LaVergne TN
LVHW071206210926
848482LV00015B/738
9780860545101